Thanasis Daradoumis, Santi Caballé, Joan Manuel Marquès, and Fatos Xhafa (Eds.)

Intelligent Collaborative e-Learning Systems and Applications

Studies in Computational Intelligence, Volume 246

Editor-in-Chief
Prof. Janusz Kacprzyk
Systems Research Institute
Polish Academy of Sciences
ul. Newelska 6
01-447 Warsaw
Poland
E-mail: kacprzyk@ibspan.waw.pl

Further volumes of this series can be found on our
homepage: springer.com

Vol. 224. Amandeep S. Sidhu and Tharam S. Dillon (Eds.)
Biomedical Data and Applications, 2009
ISBN 978-3-642-02192-3

Vol. 225. Danuta Zakrzewska, Ernestina Menasalvas, and
Liliana Byczkowska-Lipinska (Eds.)
Methods and Supporting Technologies for Data Analysis, 2009
ISBN 978-3-642-02195-4

Vol. 226. Ernesto Damiani, Jechang Jeong, Robert J. Howlett,
and Lakhmi C. Jain (Eds.)
*New Directions in Intelligent Interactive Multimedia Systems
and Services - 2,* 2009
ISBN 978-3-642-02936-3

Vol. 227. Jeng-Shyang Pan, Hsiang-Cheh Huang, and
Lakhmi C. Jain (Eds.)
Information Hiding and Applications, 2009
ISBN 978-3-642-02334-7

Vol. 228. Lidia Ogiela and Marek R. Ogiela
Cognitive Techniques in Visual Data Interpretation, 2009
ISBN 978-3-642-02692-8

Vol. 229. Giovanna Castellano, Lakhmi C. Jain, and
Anna Maria Fanelli (Eds.)
Web Personalization in Intelligent Environments, 2009
ISBN 978-3-642-02793-2

Vol. 230. Uday K. Chakraborty (Ed.)
*Computational Intelligence in Flow Shop and Job Shop
Scheduling,* 2009
ISBN 978-3-642-02835-9

Vol. 231. Mislav Grgic, Kresimir Delac, and
Mohammed Ghanbari (Eds.)
*Recent Advances in Multimedia Signal Processing and
Communications,* 2009
ISBN 978-3-642-02899-1

Vol. 232. Feng-Hsing Wang, Jeng-Shyang Pan, and
Lakhmi C. Jain
Innovations in Digital Watermarking Techniques, 2009
ISBN 978-3-642-03186-1

Vol. 233. Takayuki Ito, Minjie Zhang, Valentin Robu,
Shaheen Fatima, and Tokuro Matsuo (Eds.)
Advances in Agent-Based Complex Automated Negotiations,
2009
ISBN 978-3-642-03189-2

Vol. 234. Aruna Chakraborty and Amit Konar
Emotional Intelligence, 2009
ISBN 978-3-540-68606-4

Vol. 235. Reiner Onken and Axel Schulte
System-Ergonomic Design of Cognitive Automation, 2009
ISBN 978-3-642-03134-2

Vol. 236. Natalio Krasnogor, Belén Melián-Batista, José A.
Moreno-Pérez, J. Marcos Moreno-Vega, and David Pelta
(Eds.)
*Nature Inspired Cooperative Strategies for Optimization
(NICSO 2008),* 2009
ISBN 978-3-642-03210-3

Vol. 237. George A. Papadopoulos and Costin Badica (Eds.)
Intelligent Distributed Computing III, 2009
ISBN 978-3-642-03213-4

Vol. 238. Li Niu, Jie Lu, and Guangquan Zhang
Cognition-Driven Decision Support for Business Intelligence,
2009
ISBN 978-3-642-03207-3

Vol. 239. Zong Woo Geem (Ed.)
*Harmony Search Algorithms for Structural Design
Optimization,* 2009
ISBN 978-3-642-03449-7

Vol. 240. Dimitri Plemenos and Georgios Miaoulis (Eds.)
Intelligent Computer Graphics 2009, 2009
ISBN 978-3-642-03451-0

Vol. 241. János Fodor and Janusz Kacprzyk (Eds.)
Aspects of Soft Computing, Intelligent Robotics and Control,
2009
ISBN 978-3-642-03632-3

Vol. 242. Carlos A. Coello Coello, Satchidananda Dehuri, and
Susmita Ghosh (Eds.)
*Swarm Intelligence for Multi-objective Problems in Data
Mining,* 2009
ISBN 978-3-642-03624-8

Vol. 243. Imre J. Rudas, János Fodor, and
Janusz Kacprzyk (Eds.)
Towards Intelligent Engineering and Information Technology,
2009
ISBN 978-3-642-03736-8

Vol. 244. Ngoc Thanh Nguyen, Rados law Piotr Katarzyniak,
and Adam Janiak (Eds.)
New Challenges in Computational Collective Intelligence,
2009
ISBN 978-3-642-03957-7

Vol. 245. Oleg Okun and Giorgio Valentini (Eds.)
*Applications of Supervised and Unsupervised Ensemble
Methods,* 2009
ISBN 978-3-642-03998-0

Vol. 246. Thanasis Daradoumis, Santi Caballé,
Joan Manuel Marquès, and Fatos Xhafa (Eds.)
*Intelligent Collaborative e-Learning Systems and
Applications,* 2009
ISBN 978-3-642-04000-9

Thanasis Daradoumis, Santi Caballé,
Joan Manuel Marquès, and Fatos Xhafa (Eds.)

Intelligent Collaborative e-Learning Systems and Applications

Thanasis Daradoumis
Open University of Catalonia
Department of Computer Sciences
Multimedia and Telecommunications
Rbla. Poblenou. 156
08018 Barcelona
Spain
E-mail: adaradoumis@uoc.edu

Joan Manuel Marquès
Open University of Catalonia
Department of Computer Sciences
Multimedia and Telecommunications
Rbla. Poblenou. 156
08018 Barcelona
Spain
E-mail: jmarquesp@uoc.edu

Santi Caballé
Open University of Catalonia
Department of Computer Sciences
Multimedia and Telecommunications
Rbla. Poblenou. 156
08018 Barcelona
Spain
E-mail: scaballe@uoc.edu

Dr. Fatos Xhafa
Polytechnic University of Catalonia
Department of Languages
and Informatic Systems
Jordi Girona Salgado 1-3
08034 Barcelona
Spain
E-mail: fatos@lsi.upc.es

ISBN 978-3-642-04000-9   e-ISBN 978-3-642-04001-6

DOI 10.1007/978-3-642-04001-6

Studies in Computational Intelligence   ISSN 1860-949X

Library of Congress Control Number: 2009933888

© 2009 Springer-Verlag Berlin Heidelberg

This work is subject to copyright. All rights are reserved, whether the whole or part of the material is concerned, specifically the rights of translation, reprinting, reuse of illustrations, recitation, broadcasting, reproduction on microfilm or in any other way, and storage in data banks. Duplication of this publication or parts thereof is permitted only under the provisions of the German Copyright Law of September 9, 1965, in its current version, and permission for use must always be obtained from Springer. Violations are liable to prosecution under the German Copyright Law.

The use of general descriptive names, registered names, trademarks, etc. in this publication does not imply, even in the absence of a specific statement, that such names are exempt from the relevant protective laws and regulations and therefore free for general use.

Typeset & Cover Design: Scientific Publishing Services Pvt. Ltd., Chennai, India.

Printed in acid-free paper

9 8 7 6 5 4 3 2 1

springer.com

Preface

Intelligent Collaborative e-Learning Systems and Applications

One of the main research focuses in Computer Supported Collaborative Learning (CSCL) investigates on the improvement of on-line collaborative learning processes through the development of collaborative e-learning systems and applications which are empowered with intelligent methods and techniques. As a result, collaborative learning systems and applications are to be more powerful and flexible and also more adaptable to the learning process and thus provide better answers to the paradigmatic principles of on-line collaborative learning. In addition, virtual learning environments should be able to deeply influence the learning process by providing precise and exhaustive information and complex knowledge to all actors of the collaboration about the real dynamics of the group activity. By means of the obtained knowledge, tutors can offer an adequate support at the appropriate time and situation and evaluate the entire learning process more objectively, as well as personalize the course and its contents to the characteristics, preferences and rhythm of each learner. From the learners' standpoint, this knowledge brings them different forms of awareness about the current state of group activity as well as helps them self-regulate their individual behavior by observing and comparing each other's performance.

Moreover, important research on on-line collaborative learning practices includes the study of infrastructures that give support to those communities of contributors whose resources are managed in a peer-to-peer fashion. On-line collaborative learning processes can be improved through the employment of large scale distributed and decentralized infrastructure and the use of heterogeneous resources contributed by either the members of the learning group themselves, academic and research institutions, or third-part companies providing resources in exchange for an economic benefit. A key issue in this context is the use of economic algorithms to guarantee the efficient allocation of the contributed resources in the infrastructure. This will allow all the actors of the learning process to complement and increase the amount of the available resources. A second objective related to the infrastructure is the management of deployed learning services and applications using the contributed resources. This will require defining new mechanisms to guarantee the availability of services and learning applications as well as to evaluate its correct behavior in learning environments. Such mechanisms must take into account several variables such as demand of services, expected availability, volatility of resources and temporality.

Finally, 'learning design'-based e-learning environments seem to be promising contexts for the design of pedagogically sound e-learning events. The key principle in learning design is that it represents the learning activities and the support activities that have to be performed by different people (learners, teachers) in the context of a unit of learning. Nevertheless, there is an absence of tools that could support teachers' attempts for 'learning design' by explicitly taking into account the development of learners' cognitive skills. In addition, there is a lack of a coherent and integrated framework that could support the design of tools that should allow teachers to be involved in learning design –especially on the design of tools that support both synchronous and asynchronous communication – and focus on the development of learners' cognitive structures and critical thinking. In any collaborative learning effort, it is very important to provide appropriate intelligent tools and methods to support effective and meaningful communication among participants – both synchronous and asynchronous discussions – so that to encourage the development of core thinking skills as well as critical and creative thinking in learners.

The merge of all these synergies represents an attractive but quite laborious challenge that will yield systems capable of providing more effective answers on how to improve and enhance the on-line collaborative learning experience.

Despite the considerable progress in this field, there are still plenty of issues to investigate on how to employ the emergent computational technologies to fully support online collaborative learning and teaching activity. Four such issues concern the employment of methodologies and tools that support, on the one hand, the development of more powerful e-collaboration settings and, on the other hand, the structuring, analysis and regulation of collaborative learning interactions; the use of models that support e-learning in collaborative work experiences; and finally, the development of distributed and decentralized infrastructures to support resource allocation and services composition in collaborative peer-to-peer or Grid-based environments.

As a result, in this book we present up-to-date research approaches for developing both conceptual frameworks and computational technologies to support functional online collaborative learning and work settings. The book aims at providing tutors and researchers of online collaborative work and learning with approaches for effective and efficient means that would assist e-learning participants in enforcing and improving their online learning and working activity at both individual and group levels. Moreover, approaches in the book appeal for providing researchers and developers of online collaborative systems with fresh and innovative ideas to extend existing e-learning platforms so that they can be used efficiently in a distributed environment where learning design and material producers, service providers, and users (either tutors, learners, or academic coordinators) share similar collaborative learning and work experiences.

Among the many features highlighted in the book, which provide a significant support to the design and development of intelligent collaborative e-learning systems, we could distinguish the following:

Learning design approaches, such as adaptation, collaboration scripting, and critical thinking. Adaptive approaches to scripted collaborative learning can be seen as methods for improving the learning interactions among the students during a learning experience. Moreover, the realization of effective synchronous and asynchronous communication can be further enhanced by tools that help form such interventions that encourage the development of core thinking skills as well as creative and critical thinking in learners.

Interaction analysis approaches. Tracking group and individual activity is very important for being aware of learners' participation and performance, group functioning, as well as for assessing individual contribution in online discussions In collaborative learning activities that span over a long period of time, supporting a collaborative activity based on an asynchronous discussion or help students and teachers establish good social interaction patterns is essential. Monitoring learners' progress, identifying potential problems within groups or students and sorting them out quickly may contribute to the successful achievement of their learning goals.

Combination of supporting media, tools and other learning elements. In real complex online collaborative learning contexts, the combination of media, such as streaming, educational software, as well as other elements, such as interactions, perspectives, learning experiences and solution strategies, may result crucial in encouraging collaboration in groups which encounter many difficulties that may originate by the technical content of the learning activity and/or by the intrinsic complex nature of the collaborative work itself.

E-learning at work. E-learning is not exclusive only in educational settings. Important learning activity can also take place at work experiences. For this reason, models that support the interoperability among organizations, as well as personalized, contextualized, effective and efficient e-learning at work are needed. Moreover, communities of practice continually emerge and proliferate in the Web. New approaches are needed for organizing and encouraging group interactions and other democratic group production situations in order to enhance and improve the construction of effective knowledge.

Resource allocation and services composition in collaborative peer-to-peer or Grid-based environments. Virtual organizations (VO) normally gather several resources and interests of their members in a way that they can lend resources to or borrow them from other VO. This results in the need of designing decentralized resource allocation systems, especially for the case of collaborative peer-to-peer environments. Also, effective semantic description model are needed to automate tasks such as discovery, matching, composition and invocation of services in collaborative learning scenarios that share a variety of network resources.

Introduction

This book consists of 11 chapters organized into four major areas: (i) *Methodologies and tools that support the development of more powerful e-collaboration settings*, (ii) *Models that support e-learning in collaborative work experiences*, (iii) *Methodologies and tools for the structuring, analysis and regulation of collaborative learning interactions*, and (iv) *Distributed and decentralized infrastructures that support resource allocation and services composition in collaborative peer-to-peer or Grid-based environments*.

Methodologies and tools that support the development of more powerful e-collaboration settings

The chapters in this area are organized as follows:

The first Chapter by Demetriadis and Karakostas provides a comprehensive introduction to adaptive collaboration scripting, examining several pedagogical and technical issues. The authors bring together and emphasize the benefits of two important learning design approaches, namely adaptation of the learning environment and scripted collaborative learning. They analyze some of the major pedagogical and technical issues related to the design and development of systems for adaptive collaboration scripting (ACS systems) at two levels: at the pedagogical level, they provide methodological steps on how to build systems for adaptive collaboration scripting; at the technical level, they present a generalized architecture for systems that adaptively support learners during scripted collaboration sessions.

Constantinou et al., in Chapter 2, present an approach that combines streaming media and collaborative elements to support lifelong learning. In particular, they describe the educational value of streaming media combined with asynchronous collaborative learning and then describe the particular characteristics and limitations of the supporting Asynchronous Multimedia Collaborative Systems (AMCL). They show the new trends in AMCL systems that come to address the limitations of the current systems and to enhance the educational value of streaming media. The chapter provides ideas for educators who employ educational technologies to have the challenge of creating systems like AMCL that will make students become active consumers of digital audiovisual learning content.

Chapter 3, by Kordaki, proposes a learning framework - the MULTIPLES framework (Multiple: Learning Tools, Interactions, Perspectives, Learning Experiences and Solution Strategies) - that can be used in e-collaboration settings to encourage the development of multiple perspectives for each individual student regarding the learning concepts in focus. This framework has been designed within the context of social and constructivist learning theories, acknowledging the role of asking learners, working in both groups and individually, to face appropriately designed learning tasks by using multiple learning tools and representation systems while at the same time performing various interactions in order to produce multiple solution strategies 'in as many ways as possible'. To this end, a case study is reported that illuminates the role played by MULTIPLES in the enhancement of each individual student's views by generating different solution strategies to the tasks at hand, while at the same time expressing their inter- and intra-individual differences.

Models that support e-learning in collaborative work experiences

E-learning can be an important matter at a collaborative working environment as well. These types of environments may raise several issues that are to be addressed. Such issues are how to support the interoperability among organizations, as well as how to support personalized, contextualized, effective and efficient e-learning at work. Focusing on these issues, Capuano et al. propose a model in Chapter 4, called the Knowledge Virtual Enterprise model, where the Virtual Enterprise vision is extended with Knowledge-based assets. This model offers several advantages which are best explained by defining some real-world business scenarios, to be executed within the context of a Knowledge Virtual Enterprise instance. The scenarios are based on the idea that several organizations could put together their competences, human resources, expertise, technologies, etc. to carry out complex project activities, requiring re-sources that are usually difficult to be found in a single organization.

Thaw et al. suggest that democratically organized computer-supported cooperative work (CSCW) systems provide an ideal platform for organizing and encouraging group interactions in e-Learning and other democratic group production situations where physical meetings are not feasible. To this end, they present a system in Chapter 5, called Communities of Practice Environment (CoPE), which generated several findings. The development of workflows, for example – a process of abstracting system concepts into plain language – provides substantial benefit to non-technical users. Their experience with the system, however, revealed that while the potential for benefit is substantial, the benefit will best (if not only) manifest when the system is engaged by users who have a need for the type of democratic collaboration the system is designed to facilitate.

Methodologies and tools for the structuring, analysis and regulation of collaborative learning interactions

This area is being introduced by chapter 6 (Kordaki and Daradoumis) which presents critical thinking as a framework for structuring synchronous and asynchronous communication within learning design-based e-learning systems. The successful development of the learners' core cognitive skills and critical thinking which is strongly related to an effective discussion structuring can be achieved by the design of specific tools which can be used by both teachers and students. To that end, the authors propose a Cognitive Skill-based Communication Wizard (CSC-Wizard) for helping discussion participants formulate appropriate interventions that express their intentions more clearly and thus facilitate the development of their cognitive skills more adequately. The design of this CSC-Wizard is based on modern social and constructivist views of learning and dialogue modeling. The idea, the rationale, the architecture and the interface associated with the proposed CSC-Wizard is presented through implementing a specific example within LAMS and MOODLE systems.

Chapter 7, by Casillas and Daradoumis, goes a step further in the modeling of collaborative conversations by proposing a multi-agent system for discovering the meaning over natural-language collaborative conversations. The analysis of natural language dialogues is a complex task due to the diversity of meaning for words and expressions according to the context. This work aims at presenting a multi-agent approach for dealing with the problem of discovering the meaning of expressions written in Spanish, based on a flexible recovery system and Bayesian principles. At a first stage, agents are supposed to identify the role of the words composing a sentence. At a second stage, a second set of agents is ready to coordinate among them in order to assemble a meaning. This research sets the basis for developing an intelligent analysis tool of collaborative conversations that take place among participants in a web-based collaborative learning environment with the aim to identify participants' intentions and assess individual contribution in an automated way.

Continuing with the focus of the research on the analysis of collaborative activity based on asynchronous discussions, Caballé et. al., in Chapter 8, propose the design of a different tool, called CoLPE (Communities of Learning Practice Environment), which is actually an extension of the CoPE system presented in Chapter 5, since it extends it from the cooperative work domain to the collaborative e-learning one. The new system gives more emphasis on the development of advanced mechanisms of information management of group activity. The extracted information is processed, analyzed and provides the tutor with effective knowledge on learners' interaction behavior. This enables the tutor to adequately regulate the learning process as well as to enhance the learning group's participation by means of providing appropriate awareness and feedback. Evaluation of the approach followed showed promising results in supporting both formal and informal discussion processes that occur in current communities of learning practice.

Interaction patterns that take place in collaborative learning depend on the roles assumed by participants in the learning process. In real computer-based collaborative practices, teachers need support to be able to detect these emergent roles and undesired interaction patterns. Marcos et al. address the issue of detecting and solving negative situations in real CSCL experiences in Chapter 9, proposing a role-based interaction analysis approach. This approach is supported by a tool called Role-AdaptIA which is used to detect and help solve problematic situations in authentic computer supported collaborative learning (CSCL) scenarios. Role-AdaptIA is an adaptive interaction analysis tool based on a theoretical framework for the description of roles. The framework permits to define and characterize the roles to be taken into account in a given situation. Based on this information, Role-AdaptIA automatically detects role changes during the development of the collaborative experiences and warns the teacher about these changes. With this advice, the teacher is able to regulate the collaboration, providing support to the students to improve their interaction patterns.

Distributed and decentralized infrastructures that support resource allocation and services composition in collaborative peer-to-peer or Grid-based environments

The last two chapters raise important issues related to resource allocation and services composition in collaborative peer-to-peer or Grid-based environments. They emphasize the role that distributed and decentralized infrastructures play in these situations. Thus, in the first case, Chapter 10, by Vilajosana et al., deal with virtual organizations (VO) that gather the resources and interests of their members in a way that they can lend resources to or borrow them from other VO. To face this issue effectively, they propose DyMRA, a decentralized resource allocation system based on markets that allows inter-VO resource allocation. DyMRA is specially designed for collaborative peer-to-peer environments, where the autonomy of participants is important. The key aspect of DyMRA is that of market decentralization, that allows allocations of resources amongst different VO in spite of markets' failures. This chapter shows how the architecture of such a system is built as well as the main tools that have been developed to make DyMRA a reality. It also shows how the economy can be used to regulate the usage of resources as well as the implications of using different market mechanisms. A thorough evaluation of the system is finally provided.

The last Chapter 11, by Gutiérrez and Jorba, faces the use of distributed technologies in collaborative environments from the Grid perspective. The Learning Grid has been a quite hot issue during the last years. It is based on a secure, flexible and coordinated form of sharing network resources which are dynamically collected by individuals and institutions, and establishing mechanisms for the correct exchange of information and a strict control of the resources to share. Learning services are fundamental components of learning Grid representing functionalities that can be easily reused without knowing the details of how services have been implemented. Semantic modeling of web services promises to automate tasks such as discovery, matching, composition and invocation of services. The objective of this chapter is to present an overview of a work related with the analysis, design and implementation of semantic models for the description of learning services and their incorporation inside Collaborative Learning Scenarios based on Grid technologies.

Final Words

Intelligent Collaborative e-Learning Systems and Applications is a major research theme in CSCL and CSCW research community. It comprises a variety of research topics that focus on developing systems that are more powerful and flexible and also more adaptable to the learning process and thus provide better answers to the paradigmatic principles of on-line collaborative learning and work. The chapters collected in this book provide new insights, findings and approaches both on the analysis and the development of more powerful e-collaboration settings. Researchers will find in this book the latest trends in these research topics. On the other hand, academics will find practical insights on how to use conceptual and experimental approaches in their daily tasks. Finally, developers from CSCL community can be inspired and put in practice the proposed models and evaluate them for the specific purposes of their own work and context.

We hope the readers will find this book useful and share with us the joy!

July 2009
Editors of the Book
Barcelona, Spain

Contents

Introduction to Adaptive Collaboration Scripting: Pedagogical and Technical Issues 1
Stavros Demetriadis, Anastasios Karakostas

Combining Streaming Media and Collaborative Elements to Support Lifelong Learning 19
Charalambos Constantinou, Symeon Retalis, George Papadopoulos, Vrasidas Charalambos

'MULTIPLES': A Challenging Learning Framework for the Generation of Multiple Perspectives within e-Collaboration Settings 37
Maria Kordaki

E-Learning at Work in the Knowledge Virtual Enterprise 53
Nicola Capuano, Sergio Miranda, Francesco Orciuoli

Communities of Practice Environment (CoPE): Democratic CSCW for Group Production and E-Learning 65
David Thaw, Jerome Feldman, Joseph Li, Santi Caballe

Critical Thinking as a Framework for Structuring Synchronous and Asynchronous Communication within Learning Design-Based E-Learning Systems 83
Maria Kordaki, Thanasis Daradoumis

Constructing a Multi-agent System for Discovering the Meaning over Natural-Language Collaborative Conversations 99
Luis Casillas, Thanasis Daradoumis

CoLPE: Support for Communities of Learning Practice by
the Effective Embedding of Information and Knowledge
about Group Activity.. 113
Santi Caballe, Jerome Feldman, David Thaw

Detecting and Solving Negative Situations in Real CSCL
Experiences with a Role-Based Interaction Analysis
Approach.. 129
*José Antonio Marcos-García, Alejandra Martínez-Monés,
Yannis Dimitriadis, Rocío Anguita-Martínez, Inés Ruiz-Requies,
Bartolomé Rubia-Avi*

DyMRA: A Decentralized Resource Allocation Framework
for Collaborative Learning Environments 147
*Xavier Vilajosana, Daniel Lázaro, Joan Manuel Marquès,
Angel A. Juan*

A Semantic Description Model for the Development
and Evaluation of Grid-Based, Innovative, Ubiquous and
Pervasive Collaborative Learning Scenarios 171
Gustavo Gutiérrez-Carreón, Josep Jorba

Author Index... 189

Introduction to Adaptive Collaboration Scripting

Pedagogical and Technical Issues

Stavros Demetriadis and Anastasios Karakostas

Aristotle University of Thessaloniki, Greece
P.O. Box 114, 54124, Thessaloniki, Greece
{sdemetri, akarakos}@csd.auth.gr

Abstract. Adaptive collaboration scripting is the idea that computer-supported collaboration scripts can be adapted during run time in several of their aspects, to provide learning experiences tailored to individual and group characteristics. The pedagogical rationale of this idea is to bring together and reap the benefits of two important learning design approaches, namely adaptation of the learning environment and scripted collaborative learning. In this chapter we analyze some of the major pedagogical and technical issues related to the design and development of systems for adaptive collaboration scripting (ACS systems). At pedagogical level, we provide methodological steps on how to build systems for adaptive collaboration scripting, explaining why it is important to distinguish between "intrinsic" and "extrinsic" aspects of the script. At technical level we present a generalized architecture for systems that adaptively support learners during scripted collaboration sessions. The implementation of this conceptual framework is further illustrated by a case study on the design of a web-based system for supporting the adaptive operation of a "pyramid" type collaboration script.

1 Introduction

The idea of "script" was originally born in the psychology milieu denoting a generic pattern of behavior acquired through many years of social life [1]. It was then transferred to the collaborative learning context where by "collaboration scripts" one refers to didactic scenarios that "facilitate social and cognitive processes of collaborative learning by shaping the way learners interact with each other" [2]. A script provides specific instructions...

> "...for small groups of learners on what activities need to be executed, when they need to be executed, and by whom they need to be executed in order to foster individual knowledge acquisition" [3].

The need for using scripts emerges from the fact that collaborative learning is a complex process where it is very difficult – if not impossible – for the instructor to consider all interacting parameters in order to foster productive learning experiences [4]. Instead, it is suggested that the instructor guides the learners' interactions within the group, by implementing an appropriate collaboration script [5]. In this way one increases the probability of productive student-student and student-teacher learning interactions. Indeed, scripting collaborative learning has commonly been reported to result in improved learning outcomes [6], [7], [8].

Lately, the considerable interest that the scripting approach has gained in the computer-supported collaborative learning (CSCL) community, has motivated efforts for the formalization of collaboration scripts and the development of computer-based environments for supporting scripted collaborative learning [2], [9], [10]. At the same time, however, computer-based scripting has also been criticized for its loss of flexibility (difficulty of modifying a script in real time according to the needs of the instructional situation) [11], and also the danger of "over-scripting" collaborative activity (the pitfall of overemphasizing script imposed interactions and constraining natural collaboration) [12]. Dillenbourg and Tchounikine [13] have already emphasized the need for designers to differentiate between flexibility loss that is due to pedagogical design and undesired constraints of computer-based scripting techniques.

Against this background, we argue that building computer-based systems for adaptive scripting can be beneficial for learners and instructors in collaborative learning situations. Such systems can adaptively adjust various parameters of the collaborative process, tailoring the learning experience to the needs and characteristics of the learners. The expected benefits from such an endeavor are:

1. *Enhance peer interaction*: Implementing adaptive techniques when appropriate (such as, for example, in group formation process) is expected to establish conditions of productive peer interaction.
2. *Flexible scripting*: Adjusting the level of scripting is expected to help avoiding conditions of over- or under-scripting (depending on learner and group characteristics).

In the following, we (a) provide a theoretical background analyzing key themes of both the "adaptivity in learning design" and the CSCL approach, (b) introduce the pedagogical rationale of the adaptive collaboration scripting approach, (c) present a general architecture for building technology-environments that adaptively support scripted collaboration, and (d) illustrate how this background guided the design of a web-based environment for supporting the adaptive implementation of a "pyramid" type collaboration script.

2 Adaptivity and Collaboration

Adaptivity and collaboration are two major pillars in our current understanding on how to design effective technology-enhanced learning environments (TELEs). Adaptivity refers to the idea that TELEs can automatically adapt their operation in various aspects in order to accommodate the learning needs and preferences of the learner. Such systems are usually referred to as "Adaptive Educational Hypermedia Systems" [14] or – when accessible on the Web - "Adaptive Web-based Educational Systems (AWBES) [15].

Collaboration, on the other hand, is the idea that learning can more fruitfully take place in collaborative conditions when group members interact in order to achieve their learning objectives. When learners are supported in their collaboration using appropriately designed computer-based tools for communication and collaboration then we talk of "Computer-Supported Collaborative Learning" (CSCL)).

Of course, adaptivity needs not to refer to collaborative learning situations, nor CSCL systems need necessarily to be adaptive. However, in the crossroad of adaptive approaches and collaborative learning, novel and promising ideas emerge on how to best benefit from linking these approaches.

In the following sections we are going to present a concise overview of adaptation in technology-enhanced learning, of computer-supported collaborative learning and also, of current efforts on implementing adaptive characteristics in CSCL systems.

2.1 Adaptivity in TELEs

Adaptive educational systems are systems which use adaptivity in an educational context. Adaptive educational systems use a model of individual user's characteristics (goals, preferences and knowledge) in order to adapt their operation to the needs of the user during student-system interaction [14]. Almost all of the adaptive educational systems are based on Web platform and nearly all the systems developed since 1996 are Web-based systems [16]. The main feature of adaptive educational systems is the user model:

> The user model is a representation of information about an individual user that is essential for an adaptive system to provide the adaptation effect, effect, i.e., to behave differently for different users. [17]

The user model can be created either from the observation of user's interactions or from requesting direct input from the users (e.g. giving users an appropriate questionnaire). The process of the user model development and upholding is called "*user modelling*". The main question in user modelling is what is being modelled. The user model can record user's knowledge of the subject that is being taught, their interests, goals and tasks, background, individual traits such as cognitive and learning styles and also the context of work.

Adaptive educational systems are able to perform several adaptive procedures. Curriculum sequencing provides student with the most suitable for him/her sequence of knowledge units to follow or learning tasks to accomplish. Furthermore they can provide intelligent analysis of student solution according to student's final answers to educational problems. Interactive solving support provides student with intelligent help during solving problem procedure. They are also able to perform adaptive presentation, meaning that the system adapts the content according to the user's model. Finally adaptive educational systems are able to provide adaptive navigation support in order to change several characteristics of system navigational capabilities. [16]

Adaptation has proven to be a very useful scaffolding method in educational settings. First, adaptation provides the opportunity of efficiently supporting much wider variety of users in contrast with non-adaptive applications and, second, it can improve the efficiency and productivity of distance learning. Research has already provided encouraging evidence on the effectiveness of adaptive approaches to support learners of different profile. For example, Triantafillou et al. [18] have shown that Adaptive Educational Systems that include accommodations for cognitive styles can improve significantly learning outcomes for students of different styles. In another case, Azevedo et al. [19] report that using adaptive scaffolding can be more effective (as compared to fixed scaffolding) in helping students to regulate their learning. Similarly Bell and Kozlowski [20] provide evidence indicating that adaptive guidance has significant results on the enhancement of the learning procedure in several aspects.

2.2 Computer-Supported Collaborative Learning and the Need for Scripted Collaboration

It is not easy to define Collaborative Learning, because of the complexity and the different meanings of the words 'collaboration' and 'learning'. Dillenbourg [21] states that…:

> The broadest definition of 'collaborative learning' is that it is a *situation* in which *two or more* people *learn* or attempt to learn something *together*.

Furthermore, he suggests that "collaborative" **concerns** essentially four aspects of learning:

- A *situation*, which can be characterized as more or less collaborative, defined by the persons that are going to collaborate and depending also on their model of collaboration.
- The type of *interactions* that would occur during the learning procedure (can also be more or less collaborative).
- The type of learning *mechanisms* (could be more collaborative oriented).
- The *effects* of collaborative learning (although difficulty to measure them).

Computer Supported Collaborative Learning (CSCL) emerges from the usage of technology in collaborative learning. As Lipponen [23] emphasizes:

> … CSCL is focused on how collaborative learning supported by technology can enhance peer interaction and work in groups, and how collaboration and technology facilitate sharing and distributing of knowledge and expertise among community members.

Koschmann [22] recognizes CSCL as an emerging paradigm of educational technology which is based on very different assumptions about the notion of learning and uses new research practises and procedures. In order to differentiate the CSCL systems from the various tools for distance learning, Dimitracopoulou and Petrou [24] suggested that CSCL systems should:

- Promote learning during users activity and interactions
- Enable collaboration among group participants during a specific activity
- Support collaboration by using specific tools and functions

Although collaborative learning aims to promote fruitful interactions among learners, nevertheless it is very difficult for the learners to engage in high quality interactions by themselves. Collaboration scripts are didactic scenarios that aim to improve and support the collaborative learning process by specifying the way in which learners interact with on another. Computer Supported Collaborative Learning Scripts (CSCL scripts) are computer-based environments that guide and structure the collaboration among users who follow the activities prescribed by a collaboration script [2]. The role of the computer-based system is both to support users with communication functionalities and to structure and constrain the students' learning activity [25].

There have been, so far, various efforts towards the specification of the collaboration scripts. Dillenbourg [26] suggested that every script can be seen as a sequence of phases and each phase has at least five attributes: (a) the kind of task that has to be performed at the specific phase, (b) the composition of the group, (c) the way that task is distributed among group members, (d) the way of interaction and communication among group members and (e) script's time duration.

Similarly, Kobbe et. al [2] proposed a specification for the collaboration scripts based on the script components and the script mechanisms. The script components section includes:

- The script participants
- The activities that the participants are engaged into
- The roles that the participants would have
- The resources that are going to be given to the participants
- The script groups

The script mechanisms describe the distributed nature of the script:

- Task distribution refers on how the script components as activities or roles are distributed across script participants
- Group formation refers to the procedure or the principles that define how the script participants are distributed among the groups
- The sequencing refers to the relation which the script has (both components and groups) with the time.

Research on collaboration scripts has distinguished so far two main levels of scripting: (a) the macro level and (b) the micro level [27]. The macro level refers to the organization and the structure of the collaborative activity, for example how to organize the group's task or the specific collaborations between group members. The collaboration scripts that are focused on this macro level of collaboration are called accordingly "macro scripts" [13]. The purpose of macro scripts is the creation of appropriate learning situations in order to foster learning interactions between the group members. In contrast the micro level refers to scripts that provide support for specific activities. The "micro scripts", therefore, (as opposed to the macro scripts) emphasize the activities of individual learners.

2.3 Adaptation Methods in Collaborative Learning

Computer supported collaborative learning (CSCL) systems have already embodied characteristics from adaptive and intelligent Web-based educational systems (AIWBES). There have been two major approaches: (a) adaptive group formation and peer help and (b) adaptive collaboration support [28]. In the former approach one finds systems that perform group formation based on users' personal features and preferences [29] or users' learning records (such as interaction style) obtained during an individual learning phase [30]. Systems, in the latter approach, implement group performance modeling based on group's both learning and social characteristics [31] in order to provide interactive support during the learning process.

The adaptive group formation research concerns two main questions: what are the group formation techniques and where exactly the group formation results are used. According to the group formation techniques, groups are formed either by combining the members' individual models or by observing, analyzing and finally measuring the quality and/or the quantity of the workgroup and use these results to form groups. Hoppe [32] proposed the parameterization of the group cognitive situations based on the members' individual models, in order to describe the initial conditions of a group. A group should be formed according to task relevant hypothesis which is based on the individual models, for example a group should be formed to work on a problem if there is at least one group member who is capable to solve the problem individually. There are other cases where the group formation phase follows an individual learning phase necessary either to construct the members' models [30], [33], [29] or to identify a specific situation that triggers the beginning of the collaboration phase, for example when the system "feels" that the learner needs a review of the material that has been presented [34].

Usually, the group formation process is used as an automatic mechanism in order to construct groups either in the beginning of the collaboration process or for reforming groups between the collaboration activities. Peer helping can be considered as a subcategory of group formation, since the system is trying to locate the most suitable group member to accomplish a task with a specific partner. So, "peer help" can be seen as a group formation technique for dyads based on the members' individual models [35], [36].

Adaptive collaboration support concerns the aspects of users and groups (and their activities) that have to be modeled and can be inferred or observed in the user/user or system/user interaction in order to support collaboration. For example the group performance models can contain both learning related aspects (what is the student domain knowledge, what type of exercises does the student prefer, what mistakes has the student made) and also social aspects (the members' motivation or the members' participation) [31]. In some other cases the system identifies specific actions [37] or circumstances during the collaboration in order to adapt the task presentation [38] or to enhance the collaboration process by proposing specific actions [39]. Adaptive collaboration support also refers to the adaptation techniques and methods that are tailored to the specific needs of collaborative learning and to the language that has to be used in order to describe both collaboration activities and adaptation methods.

Nevertheless, no explicit effort so far has been reported for embodying adaptation techniques in systems for scripted collaboration. Gweon et al. [40] provide research-based evidence in favor of the effectiveness of adaptive scripting when learning in an on-line collaborative environment, however, these authors use the term "scripting" to refer to the provision of support to collaborative students in the form of prompts.

In the following section we discuss the issue of implementing adaptive techniques in systems for computer supported collaboration scripting. As we shall demonstrate this is not an issue of simply transferring adaptation methods in the context of a collaboration script as one major question is always what aspects of a script can be appropriate candidates for adaptation.

3 Adaptive Collaboration Scripting

Adaptive collaboration scripting aims to implement collaboration scripts in an adaptive mode, meaning that the technology-based systems used to support the activity can implement various adaptation methods during the scripted collaboration so that the learners maximize the benefits from the learning experience.

We call such systems "systems for adaptive collaboration scripting" or ACS systems and, in general, we suggest that the adaptation methods in ACS systems should focus on two major objectives: (a) *enhance peer interaction*, and (b) establish conditions of *flexible scripting*.

(a) Enhancing Peer Interaction: Implementing user modelling techniques and respective adaptation rules, when appropriate, can help establish conditions of more effective peer interactions. For example, when recording learners' profile during group formation process to create hetero- or homogenous groups. Although scripting in itself is already an effort to guide and trigger peer interactions, if the learners' profile (and consequently the group profile) does not favour the intended peer interaction then the scripted collaborative experience might be uncreative and trivial. To avoid such situations it is advisable to investigate early in the beginning of the script what might be an appropriate group formation method, taking into account the learners' profile in relation to the planned peer interactions. This is the case, for example, when implementing the Arguegraph script [41]. The core mechanism of learning in this script is the argumentation that takes place between group partners. Accordingly, the script begins by recording (through an appropriate questionnaire) the learners' individual profiles regarding their views and opinions on the issue under study. Next, pairs are formed so that students with considerably different opinions are in the same pair and, consequently, argumentation can be based on fertile ground.

From this perspective, therefore, "adaptive scripting" refers to the ability of an ACS system to adapt various conditions of collaborative work (such as the group synthesis) so that that the possibility for productive peer interactions is increased. As illustrated from the Argugraph example, to attain this it is necessary to apply user modelling methods that help create groups with the desired heterogeneity or homogeneity level.

(b) **Flexible Scripting:** Adjusting the level of scripting to avoid conditions of over- or under-scripting is also an important objective of the adaptive collaboration scripting approach. This essentially means that when learners collaborate following the instructions of a script, they should not feel restrained by excessive script guidance (over-scripting) not be left unsupported when they need more guidance and scaffolding during the collaboration activity (under-scripting). Flexible scripting, should adjust the level of support and guidance at an optimum level considering the learners' needs and characteristics. During group activity it is important that the system "observes" the performance of the group members and adaptively offer (or remove) support depending on the value of selected parameters that model group performance.

To proceed towards this direction we first need a framework for the specification of collaboration scripts that will help us analyze the structural components of a script. Kobbe et al. [2] argue that without defining a script independent from its particular implementation in a computer-supported learning environment, systematic research on collaboration scripts is difficult because one cannot differentiate between effects from the script and effects from its particular computational implementation.

As already said, every script can be seen as a sequence of phases with five major attributes each: (a) the kind of task that has to be performed at the specific phase, (b) the composition of the group, (c) the way that task is distributed among group members, (d) the way of interaction and communication among group members and (e) script's time duration [16]. Already in listing the above script features one can identify aspects for potential adaptation. For example, the way that the task is distributed among group members can be adapted to fit learners' background.

However, the idea that a script is a didactic scenario that *must* be followed raises an interesting question in relation to adaptation: what can be adapted and under what conditions? If adaptation means changing the conditions of the learning experience to accommodate the learner, then to what extend should the script imposed constraints be followed "as is" or adapted according to a specific adaptation model? The answer that we offer is that instructional design should seek to optimize the flexibility of the learning activity without however loss of its cohesiveness [42]. In other words, the learning design should introduce elements that can be flexibly changed/adapted during the activity but without reducing the overall value of the learning experience which is strongly connected to the constraints and demands imposed to learners (cohesiveness) in order to individually or collaboratively process information.

Not any script feature, therefore, can be a candidate for adaptation. The core features of a script, those that give to the script its specific pedagogical character and value, can not be changed (or adapted) in any way. Dillenbourg and Tchounikine [13] call these features *"intrinsic"* constraints and emphasize that these set up the limits of flexibility, that is, they define what can not be changed within the scripted learning experience if the pedagogical purpose of the script is to remain intact. By contrast, *"extrinsic"* constraints refer to script's aspects that can be changed/adapted in order to provide room for learners to adapt the learning experience to their own preferences and characteristics.

We further suggest that extrinsic constraints can be seen as belonging to either of two categories: (a) *"Non-pedagogical"*, that is constraints without any pedagogical relevance. These constraints can be altered by the teacher and/or the students simply to make the script to better accommodate the conditions of the specific implementation (for example, extending the duration of a phase because of a learner's temporal inability to meet a deadline). (b) *"Pedagogical"* constraints that can (should) be adapted in order to provide a well suited learning experience to the specific group of learners (for example, repeating a phase of the script to help novice learners better understand the material or the collaborative process).

From this perspective therefore, adaptive collaboration scripting is based on representations of scripts where it is possible to define intrinsic and extrinsic (also pedagogical and non-pedagogical) characteristics. In the case of extrinsic constraints (that can be adapted), the representation should include also appropriate adaptation models (criteria and rules) for implementing any permitted user/group based adaptations.

4 A Proposed System Architecture

A proposed architecture for building systems that support adaptive collaboration scripting is shown in Figure 1. The architecture consists of three main layers: the *storage* layer which includes all the necessary elements to perform adaptation, the *group management* layer which constructs groups and manages group characteristics and, finally, the *runtime* layer which is the closest layer to the learner and contains all the interactions with him/her.

The storage layer primarily consists of the *learner/group model* component, which includes learner's cognitive characteristics and preferences about the learning process or the domain. It also records the learner/group performance during the educational procedure and the actual learner/group knowledge space.

Learner/group model is input to the *adaptation policy* component which actually provides the representation of the script pedagogical extrinsic features and enables system users to activate adaptations on any of these aspects, by editing a respective *adaptation profile*. Adaptation profiles (there should be at least one) can be conceptualized as a set of choices predefined by the tutor, on the adaptation rules that the system should activate and the aspects of the user/group model that should be considered. Adaptation policy communicates with the *adaptation model* component which stores the rules for describing the kind of adaptation that is activated during system runtime. For example content selection rules define the procedure for selecting appropriate resources that should be presented to learners.

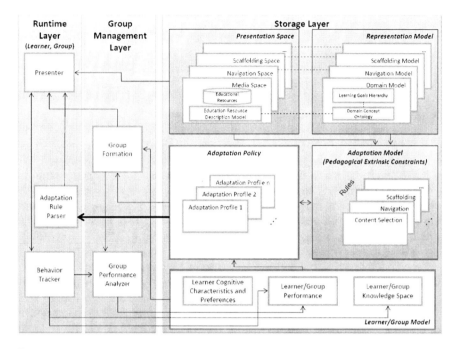

Fig. 1. Architecture of Adaptive Collaborating Scripting Systems (adapted from [43])

Similarly, scaffolding rules define the process of adaptively providing to learners on task supportive feedback.

Presentation Space contains the respective presentation modules, that is, all necessary information to be presented onscreen when a specific adaptation rule is enacted. For example media space contains all forms of learning material that can be presented by the system when performing content adaptation. Each module in the presentation space (media, navigation, scaffolding, etc.) has its counterpart in the *representation model*, which provides a representation of its structure, relevant to a learning goal hierarchy. If, for example, the system identifies three levels of learner/group competence (novice, advance, expert) the relationship between these levels is represented in the scaffolding model (within the representation model component).

The *group management layer* consists of the *group formation* and the *group performance analyzer* modules. Group formation module defines groups of students, based on available information from the user/group model and the active adaptation profile. Adaptation profile is responsible for providing rules for group formation (for example forming homogeneous groups of three based on their prior knowledge). Group performance analyzer analyzes the performance of the group and sends the filtered data to group performance module of user/group model. It has also as input the group structure from the group formation module in order to identify the type of group it analyzes.

Finally, the *runtime layer* consists of a behaviour tracker, an adaptation rule parser and the presenter. The behaviour tracker records learner actions and informs the learner performance and the learner knowledge space in the user/group model. During teamwork the tracker sends the data first to the group performance analyzer in order to produce a representation of the teamwork depending on all team members' contributions. Adaptation rule parser interprets adaptation policy output in order to give specific instructions to the presenter on how to display adaptive material onscreen. The presenter accordingly searches for the relevant material from the appropriate presentation space.

5 A Design Case Study: The Pyramid Script

5.1 What Is the "Pyramid" Script?

Generally in a learning activity of the pyramid-type each participant works first individually studying a problem (or any learning material) and then participates in a workgroup of a gradually increasing size, to collaboratively process the material from a certain perspective [9], [44].

Using the notion of the "pyramid" we designed a collaboration script to support students when engaged in case-based learning (CBL). The objective of our "pyramid" script is to help students develop a deep conceptual understanding of a complex domain by collaboratively analysing relevant cases. The script engages the participants in the processing of several cases, guiding their collaboration in groups of various sizes and with the use of various communication media.

The "pyramid" script for CBL consists of four phases:

(a) An individual study of one or more cases
(b) A collaboration phase in a small group
(c) A collaboration phase in a larger group communicating through asynchronous discussion tools, and
(d) A debriefing phase conducted in the classroom ("face-to-face" mode).

In the first phase students are assigned to study individually one or two representative domain cases, supported by instructor's guidelines and producing a report deliverable where they analyze the case(s). In the second phase students form small groups (with 2 or 3 members) and are supported by the instructor to discuss and collaboratively produce a synthesis, based on the cases that they individually studied in the first phase. In the third phase, students form larger groups (with 6 or more participants). Each small group (of previous phase 2) is represented in a larger group by one of its members. Students in these large groups are provided with new case-based material and are guided to asynchronously discuss specific domain issues (different assignment for each large group). The objective for the students in each group is to produce a report deliverable stating their final conclusions on the key issues they have been assigned. In the last phase the instructor presents in the classroom the main conclusions of the workgroups, highlighting the issues addressed satisfactorily by collaborating students and also emphasizing issues that were – perhaps – overlooked by them.

We implemented the pyramid script in a postgraduate course on technology-supported pedagogical innovation in secondary education. In this implementation we used the Blackboard LMS to support students from distance and provide an asynchronous discussion board for them to collaborate in the third phase. The research results [45] indicated that students consider the script to be an apt and effective method for case-based learning, although they ask for extensive instructor's support on how to collaborate, especially during the third phase where they communicate asynchronously. Some students also ask for more time to accomplish the activity in this phase.

Based on these results that suggest adaptation to respond to students' learning needs, we advance our research towards developing an adaptive web-based system for implementing pyramid scripts. Table 1 presents the intrinsic and extrinsic

Table 1. "Pyramid" script intrinsic and extrinsic constraints

Flexibility points	Intrinsic (in-) vs. Extrinsic (ex-) pole
Script edition	
Refers to changing activities or their order	Number of group members is growing after each phase (in-)
	Group deliverable at the end of each phase (in-)
	Sequence of phases (ex-), e.g. some phase can be repeated
Script instantiation	
Refers to changing contents	Use only cases as learning material (in-)
	The amount of learning material is not fixed (ex-), e.g. the tutor is able to provide learners with extra cases
Session set up	
Refers to defining groups	Small group formation (ex-), e.g. group formation in second phase may vary depending on student models
	Larger group formation (ex-), e.g. group formation may depend on group performance model (during activities in second phase)
Refers to defining time frame	Script phases have to follow immediately one another (in-)
	Time frame of the phases can be flexible (ex-), e.g. the tutor is able to expand phases' duration if there are specific needs
Run time	
Refers to changing groups	The group formation can change (ex-), e.g. tutor has the option to interfere in group formation
Refers to changing deadlines	The report of the deliverables, in all the three first phases, has to be submitted by all learners/groups in order to continue (in-)

script features as defined in our current design approach. These features are organized in four categories, as suggested by Dillenbourg and Tchounikine [13]:

(a) "Script edition" which refers to the modification of the script structure
(b) "Script instantiation" which refers to script content
(c) "Session set up" which refers to scripts session parameters, and
(d) "Run time" which refers to the management of script enactment.

5.2 Adaptation Techniques in "Pyramid" Script

To design an adaptive computer-based system for implementing the pyramid script, we deem as appropriate to consider at least two adaptation approaches: (a) group formation methods based on the individual student models, and (b) adaptation techniques based on group performance model. Individual student modelling is necessary for adapting the learning conditions right from the beginning of the script implementation, for example assigning appropriate level of cases to the participants in the first script-phase, depending on their prior knowledge.

We expect that by adapting the level of case-based material (for example, its complexity) we offer to novice students an entry point to the activity that keeps their cognitive load at manageable level.

In typical "pyramid" script implementations small groups (in the second phase) are formed randomly. We suggest that group formation techniques can be used at this point to enable adaptive group formation based on pedagogically sound hypotheses. For example, groups can be formed with the participants having studied contradictory (or complementary) material in the first script-phase, so that the synthesis work they are engaged into during the second phase will result in fruitful discussion interactions.

Furthermore, in the second script-phase techniques of group performance modelling can be used to keep track of various group performance parameters. The group performance model could have both learning and social aspects, such as measuring group members' participation and motivation. Based on this group model the system can take decisions (criteria control) on the adaptation rules that should be enacted. If a criteria control is not satisfied (i.e., group members participation is at low level) the system might automatically activate an adaptation rule (for example, activating a scaffolding/motivating mechanism) (automatic adaptation) or inform the teacher to take any appropriate measures (semi-automatic adaptation).

Finally, the group formation in the third phase (larger groups) can be based on the group models that were created in the second phase. For example, larger groups could be formed depending on the quality of students' deliverable in the second phase, so that the larger groups could be supported accordingly with additional help and guidance. The criteria control and the adaptation mechanisms that were implemented in the second phase could also be applied in the third phase.

In Figure 2, we present the UML activity diagram of the whole "pyramid" script with the general adaptation features that have been discussed. The diagram

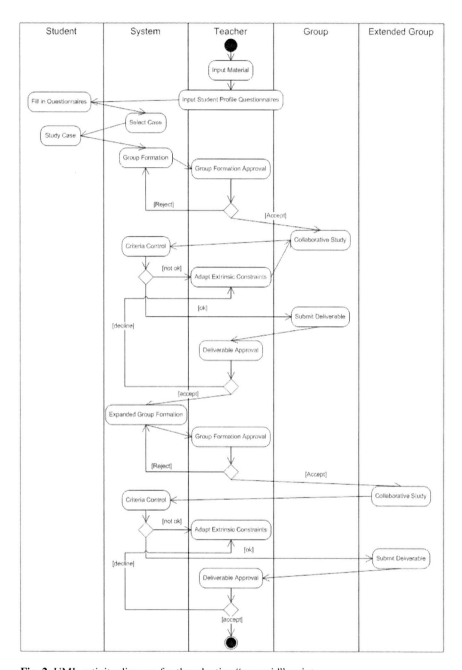

Fig. 2. UML activity diagram for the adaptive "pyramid" script

also highlights the time flow of the script and the main roles that get involved into the whole procedure.

6 Conclusions and Future Work

In this work we have argued in favor of an adaptive approach to scripted collaborative learning as a method for enhancing the learning interactions among the students during the learning experience. We have suggested that for building systems for adaptive scripting one need first to distinguish between "intrinsic" and "extrinsic" script constraints and develop a respective computerized script representation. Based on that, the script system can "decide" when to enact adaptation interventions, which can be automated (the relevant adaptation model is build into the system as a whole) or semi-automated (the system simply informs the instructor on the situation and he/she then decides on the kind of intervention/adaptation that is appropriate).

We have also presented a design case study, where the adaptive approach is implemented in the case of a "pyramid" collaboration script, suggesting two major adaptation mechanisms: (a) group formation techniques based on student individual models, and (b) adaptation techniques based on group performance model.

Future work includes the development of a system for supporting students in pyramid-type scripted collaboration. In building the system, we identify the script intrinsic/extrinsic constraints and then the exact adaptation rules that will be integrated to the system. In evaluating the system we will focus on assessing the students' satisfaction and quality of learning compared to the non-adaptive treatment and also the cost/efficiency ratio of using the system from the instructor's viewpoint.

References

1. Schank, R.C., Abelson, R.: Scripts, Plans, Goals, and Understanding. Erlbaum Association, Hillsdale (1977)
2. Kobbe, L., Weinberger, A., Dillenbourg, P., Harrer, A., Hämäläinen, R., Fischer, F.: Specifying computer-supported collaboration scripts. International Journal of Computer-Supported Collaborative Learning 2(2-3), 211–224 (2007)
3. Weinberger, A., Stegmann, K., Fischer, F., Mandl, H.: Scripting argumentative knowledge construction in computer-supported learning environments. In: Fischer, F., Kollar, I., Mandl, H., Haake, J. (eds.) Scripting computer-supported collaborative learning: Cognitive, computational and educational perspectives, pp. 191–211. Springer, New York (2007)
4. Dillenbourg, P., Baker, M., Blaye, A., O'Malley, C.: The evolution of research on collaborative learning. In: Spada, E., Reiman, P. (eds.) Learning Human and Machine: Towards an interdisciplinary learning science, pp. 189–211. Elsevier, Oxford (1995)
5. O'Donnell, A.M., Dansereau, D.F.: Scripted cooperation in student dyads: A method for analyzing and enhancing academic learning and performance. In: Hertz-Lazarowitz, R., Miller, N. (eds.) Interaction in cooperative groups: The theoretical anatomy of group learning, pp. 120–141. Cambridge University Press, London (1992)

6. Rummel, N., Spada, H.: Can people learn computer-mediated collaboration by following a script? In: Fischer, F., Kollar, I., Mandl, H., Haake, J. (eds.) Scripting computer-supported collaborative learning: Cognitive, computational and educational perspectives, pp. 39–55. Springer, New York (2007)
7. Kollar, I., Fischer, F., Slotta, J.D.: Internal and external collaboration scripts in web-based science learning at schools. In: Koschmann, T., Suthers, D., Chan, T.W. (eds.) Computer supported collaborative learning 2005: The next 10 Years, pp. 331–340. Lawrence Erlbaum, Mahwah (2005)
8. Weinberger, A., Fischer, F., Mandl, H.: Fostering computer supported collaborative learning with cooperation scripts and scaffolds. In: Stahl, G. (ed.) Computer Support for Collaborative Learning: Foundations for a CSCL Community. Proceedings of the Conference on Computer Support for Collaborative Learning, pp. 573–574. Erlbaum, Hillsdale (2002)
9. Turani, A., Calvo, R.: The Potential Use of Collaboration Scripts in Synchronous Collaborative Learning. In: Proceedings of IMCL2007 Conference, Amman, Jordan, April 18-20 (2007)
10. Bote-Lorenzo, M.L., Gómez-Sánchez, E., Vega-Gorgojo, G., Dimitriadis, Y.A., Asensio-Pérez, J.I., Jorrín-Abellán, I.M.: Gridcole: A tailorable grid service based system that sup-ports scripted collaborative learning. Computers & Education (accepted) (retrieved November 20, 2007), http://dx.doi.org/10.1016/j.compedu.2007.05.004
11. Dillenbourg, P., Jermann, P.: Designing interactive scripts. In: Fischer, F., Kollar, I., Mandl, H., Haake, J. (eds.) Scripting computer-supported collaborative learning: Cognitive, computational and educational perspectives, pp. 276–301. Springer, New York (2007)
12. Dillenbourg, P., Tchounikine, P.: Flexibility in macro-scripts for computer-supported collaborative learning. Journal of Computer Assisted Learning 23(1), 1–13 (2007)
13. De Bra, P., Brusilovsky, P., Houben, G.: Adaptive Hypermedia: From Systems to Framework. ACM Computing Surveys 31(4) (1999)
14. Brusilovsky, P.: Adaptive hypermedia. In: Kobsa, A. (ed.) User Modeling and User Adapted Interaction, Ten Year Anniversary Issue, vol. 11(1-2), pp. 87–110 (2001)
15. Brusilovsky, P., Peylo, C.: Adaptive and intelligent Web-based educational systems. In: Brusilovsky, P., Peylo, C. (eds.) International Journal of Artificial Intelligence in Education Special Issue on Adaptive and Intelligent Web-based Educational Systems, vol. 13(2-4), pp. 159–172 (2003)
16. Brusilovsky, P., Millán, E.: User models for adaptive hypermedia and adaptive educational systems. In: Brusilovsky, P., Kobsa, A., Nejdl, W. (eds.) Adaptive Web 2007. LNCS, vol. 4321, pp. 3–53. Springer, Heidelberg (2007)
17. Triantafillou, E., Pomportis, A., Demetriadis, S.: The design and the formative evaluation of an adaptive educational system based on cognitive styles. Computers and Education 41, 87–103 (2003)
18. Azevedo, R., Cromley, J., Winters, F., Moos, D., Greene, J.: Adaptive human scaffolding facilitates adolescents' self-regulated learning with hypermedia. Instructional science 33(5-6), 381–412 (2005)
19. Bell, B., Kozlowski, S.: Adaptive guidance: Enhancing self regulation, knowledge, and performance in technology-based training. Personnel Psychology 55, 267–306 (2002)
20. Dillenbourg, P.: What do you mean by collaborative learning? In: Dillenbourg, P. (ed.) Collaborative-learning: Cognitive and Computational Approaches. Advances in Learning and Instruction series, pp. 1–20. Pergamon, Elsevier (1999)

21. Koschmann, T.: Paradigm shifts and instructional technology. In: Koschmann, T. (ed.) CSCL: Theory and practice of an emerging paradigm, pp. 1–23. Lawrence Erlbaum Associates, Inc, Mahwah (1996)
22. Lipponen, L.: Exploring Foundations for Computer-Supported Collaborative Learning. In: Stahl, G. (ed.) 4th Computer Support for Collaborative Learning: Foundations for a CSCL Community (CSCL 2002), USA, pp. 72–81 (2002)
23. Dimitracopoulou, A., Petrou, A.: Advanced collaborativedistance learning systems for young students: Design issues and current trends on new cognitive and meta-cognitive tools. THEMES in Education International Journal (2005)
24. Tchounikine, P.: Conceptualizing CSCL Macro-Scripts Operationalization and Technological Settings. International Journal of Computer-Supported Collaborative Learning (2008) (to appear)
25. Dillenbourg, P.: Over-scripting CSCL: the risks of blending collaborative learning with instructional design. In: Kirschner, P.A. (ed.) Three Worlds of CSCL. Can We Support CSCL?, pp. 61–91. Open Universiteit Nederland, Heerlen (2002)
26. Fischer, F., Kollar, I., Mandl, H., Haake, J.: Perspectives on collaboration scripts. In: Fischer, F., Kollar, I., Mandl, H., Haake, J. (eds.) Scripting computer-supported collaborative learning: Cognitive, computational and educational perspectives, pp. 191–211. Springer, New York (2007)
27. Brusilovsky, P., Peylo, C.: Adaptive and Intelligent Web-based Educational Systems. International Journal of Artificial Intelligence in Education 13, 156–169 (2003)
28. Carro, R.M., Ortigosa, A., Schlichter, J.: Adaptive Collaborative Web-based Courses. In: Cueva, J.M., González, M., Joyanes, L., Labra, E., Paule, M.P. (eds.) Web Engineering. LNCS, pp. 130–133. Springer, Heidelberg (2003)
29. Quignard, M., Baker, M.: Favouring modellable computer-mediated argumentative dialogue in collaborative problem-solving situations. In: Proceedings of the 9th International Conference on AI in Education, pp. 129–136. IOS Press, Amsterdam (1999)
30. Vizcaino, A., Contreras, J., Favela, J., Prieto, M.: An adaptive, collaborative environment to develop good habits in programming. In: Proceedings of the 5th International Conference on Intelligent Tutoring Systems, Montreal, Canada, pp. 262–271 (2000)
31. Hoppe, U.: Use of multiple student modelling to parameterize group learning. In: Greer, J. (ed.) Artificial Intelligence in Education, Proceedings of AI-ED 1995, 7th World Conference on Artificial Intelligence in Education, Washington, DC, August 16-19, 1995, pp. 234–249 (1995)
32. Baker, M.J., de Vries, E., Lund, K.: Designing computer-mediated epistemic interactions. In: Lajoie, S.P., Vivet, M. (eds.) Proceedings of the International Conference on Artificial Intelligence and Education, Le Mans, pp. 139–146. IOS Press, Amsterdam (1999)
33. Ikeda, M., Go, S., Mizoguchi, R.: Opportunistic group formation. In: Boulay, B.d., Mizoguchi, R. (eds.) AI-ED 1997, 8th World Conference on Artificial Intelligence in Education. IOS, Amsterdam (1997)
34. Greer, J., McCalla, G., Collins, J., Kumar, V., Meagher, P., Vassileva, J.: Supporting Peer Help and Collaboration in Distributed Workplace Environments. International Journal of Artificial Intelligence in Education 9, 159–177 (1998)
35. McCalla, G.I., Greer, J.E., Kumar, V.S., Meagher, P., Collins, J.A., Tkatch, R., Parkinson, B.: A peer help system for workplace training. In: Boulay, B.d., Mizoguchi, R. (eds.) AI-ED 1997, 8th World Conference on Artificial Intelligence in Education, pp. 183–190. IOS, Amsterdam (1997)

36. Soller, A.: Supporting social interaction in an intelligent collaborative learning system. International Journal of Artificial Intelligence in Education 12(1), 40–62 (2001)
37. Muehlenbrock, M., Hoppe, U.: Computer supported interaction analysis of group problem solving. In: Proceedings of the Conference on Computer Supported Collaborative Learning, Palo Alto, CA, pp. 398–405 (1999)
38. Constantino Gonzalez, M.A., Suthers, D., Escamilla De Los Santos, J.G.: Coaching web-based collaborative learning based on problem solution differences and participation. International Journal of Artificial Intelligence in Education 13(2-4), 261–297 (2003)
39. Gweon, G., Rosé, C.P., Carey, R., Zaiss, Z.S.: Providing Support for Adaptive Scripting in an On-Line Collaborative Learning Environment. In: Proceedings of the SIGCHI conference on Human Factors in computing systems, Montréal, Québec, Canada, pp. 251–260 (2006)
40. Jermann, P., Dillenbourg, P.: Elaborating new arguments through a CSCL scenario. In: Andriessen, G., Baker, M., Suthers, D. (eds.) Arguing to Learn: Confronting Cognitions in Computer – Supported Collaborative Learning Environments. CSCL Series, pp. 205–226. Kluwer, Amsterdam (2003)
41. Demetriadis, S., Pombortsis, A.: e-Lectures for Flexible Learning: a Study on their Efficiency. Journal of Educational Technology & Society 10(2), 147–157 (2007)
42. Karampiperis, P., Sampson, D.: Adaptive Learning Resources Sequencing in Educational Hypermedia Systems. Educational Technology & Society 8(4), 128–147 (2005)
43. Hernández, D., Asensio-Pérez, J., Dimitriadis, Y.: IMS Learning Design Support for the Formalization of Collaborative Learning Patterns. In: Proceedings of the Fourth IEEE Inter-national Conference on Advanced Learning Technologies, Joensuu, Finland, pp. 350–354 (2004)
44. Demetriadis, S., Liotsios, K., Pombortis, A.: Scripted collaborative learning in educational activity of blended format: a case study. In: Proceedings of the International Conference for Open and Distance Learning, Athens, Greece, pp. 328–334 (2007) (in Greek)

Combining Streaming Media and Collaborative Elements to Support Lifelong Learning

Charalambos Constantinou[1], Symeon Retalis[2], George Papadopoulos[1], and Vrasidas Charalambos[3]

[1] Department of Computer Science, University Of Cyprus, Cyprus
 constandinou1@gmail.com, george@cs.ucy.ac.cy
[2] Department of Technology Education and Digital systems, University of Piraeus, Greece
 retal@unipi.gr
[3] School of Education, University of Nicosia, Cyprus
 pambos@cardet.org

Abstract. This chapter presents the educational value of streaming media combined with asynchronous collaborative learning and describes the particular characteristics and the limitations of the supporting Asynchronous Multimedia Collaborative Systems (AMCL). Then CELSIA, an innovative AMCL system, is described as an example of new trends in AMCL systems that come to address the limitations of the current systems and to enhance the educational value of streaming media!

1 The Use of Streaming Media in Education

According to cue summative learning method [1] the use of various channels of communication and stimuli helps in the retaining of knowledge and in the acquiring of new knowledge and skills. Audiovisual material can be a great resource for education thus the last decades we have witnessed a lot of research and various attempts to utilize it in classrooms and generally in education. Several research educational studies since the beginning of the 20th century found that films, videos and moving pictures are very helpful in: [2]

- attracting learners' attention,
- the presentation and clarification of complicated issues of a course subjects (e.g. natural phenomena via experimental demonstrations),
- the retaining of information,
- the motivation of learners for subjects connected to everyday life (e.g. news).

Bowie in an analytical review of studies that concern the educational use of films argued that audiovisual material:

- It is effective in discovery learning
- It can be used for demonstrating the solution of a problem
- It's appropriate for developing skills of attention and observation of details
- It can influence positively learner's self efficacy
- It improves creativity, imagination and aesthetics [3].

Yet beneath the apparently unproblematic appeal of streaming media, there is a counterargument which states that these media -and video in particular- is a passive educational mean which creates passive learners. The conclusion of a relevant research was that audiovisual material wasn't successful in classrooms [4], except of foreign languages.

In a first attempt to avoid passiveness, lecture material was broadcasted via TV to remote students while any questions or comments were posed to the instructor using usual telephone technology. This attempt was successful since many studies showed that students can learn as much from such broadcast lectures as from live classroom attendance [5], but suffers from a great problem. It is a synchronous model which means that everybody must meet at an appointed time and date. This model suffers from the fact that students cannot participate on-demand which means that it is more complex. Except from this, if the number of students that are watching the same lecture at the same time is big then the collaboration and as an effect the education value of the broadcasted lecture between them is reduced.

In order to avoid these problems we can use an asynchronous model of communication. In an asynchronous model students can participate on-demand and the number of students who are watching the lecture simultaneously is reduced.

This chapter is focused on the idea of *Asynchronous Multimedia Collaborative Learning* (AMCL), which is a relatively new concept although its origins can be found in a European funded Socrates ODL project, called SHARP: Shared Representation of Practice [6].

AMCL could be said to be a mixture of asynchronous collaborative learning with streaming multimedia content resources. AMCL combine the richness of multimedia representation and demonstrations of practice with the flexibility in the use of time for communication. The AMCL is a rather new education medium and philosophy, still unexplored.

The AMCL systems almost look like the web-based discussion fora, with the addition that the user:

- can post messages that are videos, audios (i.e. not only texts) which can be delivered via streaming technologies (like video on demand)
- can annotate-comment on specific "frame" of the message (e.g. when a specific term was explained).

Currently there are just few AMCL systems, most of them created for research purposes within universities. Some of them are quite difficult to use, luck of important functions or are specialized and limited in certain aspects. This is why there is a need to design and develop AMCL systems that could meet the requirements of end users.

The structure of this chapter is as follows. First we will give an overview of the reasons why we need to combine Streaming Multimedia technologies with collaborative learning techniques. Then the current state of the art in the domain of AMCL systems giving emphasis on their added value in education as well as on their limitations will be presented. Furthermore, two learning scenarios will be described in order to show the importance of the AMCL systems in education and the need to be enriched with new functionality. Based on these scenarios the new trends in AMCL systems that come to address the limitations of current AMCL systems will be analysed.

2 Combining Streaming Media with Collaborative Learning

Video is considered to be by many scientists as an instructional medium that has a great educational value especially for visual and auditory learners.

Video can create excitement, emotions, and help students to keep attention to the lesson. For example consider a group of students in a classroom that are studying china for the lesson of geography. They can view a video that describes china, listen to the Chinese language and traditional songs and "visit" great attractions in few minutes! The students will be more excited and motivated to learn and it will be easier for them to understand things that otherwise they had to imagine. In addition the video will enhance their retention since students visualize important information and transfer abstract concepts into concrete and easier to remember objects.

Several attempts to use video in education have been made especially at the 80s and 90s mostly with the use of videotapes, television digital video and CD ROMs. According to [7] the very early attempts were mostly used as part of instructional pedagogy while at the 90s we have witnessed many constructivist paradigms especially with the use of digital desktop video and the upcoming streaming video technologies that emerged at the late 90s.

The idea to combine Streaming Media (especially digital video) with collaborative learning techniques is based on the 3Is (Image, Interaction, Integration) framework shown in table 1 [7].

Table 1. The pedagogic framework 3I [Source: 8]

Value	Technology	Control
Image	Film, Television, video	Educator
Image + Interaction	Multimedia CD-R	Student
Image + Interaction + Integration	Streaming media	Student + Educator

According to table 1 when talking for Film, Television and video the added value to education is "Image". Students can benefit from the visual richness of video since it enhances attraction, aids retention and recall and is explanative when verbal forms are not enough.

In addition to this Goodyear & Steeples note that video can provide vivid descriptions to articulate tacit information and knowledge difficult to articulate through text and verbally [6]. When the value of video is "image" then the educator has the control of the teaching procedure. Film, Television and video of course are very important in education but they are very passive means since there is no interaction between students and these media.

Streaming Media have another added value except from "Image" and "Interaction". Streaming media can be linked with other supporting elements such as related videos, texts and resource links. According to the 3Is framework [7] this added value is called "Integration". It is very easy to understand that all these elements enhance the learning experience, since students can have access to more information if it is necessary. Although streaming media are excellent tools for educational purposes they are still passive educational means since there is no collaboration and communication between the educator and students.

The combination of collaborative learning and streaming media can be the answer to the communication and collaboration problem since the web enable various types of synchronous and asynchronous communication and collaboration such as discussion forums, chat and media sharing. We will focus in Asynchronous communication since it has some advantages over synchronous communication, the most important one that both Students and Educator participate in the educational process regardless of distance and time. The AMCL systems are designed to combine streaming media with asynchronous collaborative tools and so are ideal to enhance the learning experience and avoid passiveness.

It is also widely known that students do not interact among them if there is no certain reason or motivation. Thus collaborative learning must be enhanced with other activities, in order to develop interaction, information exchanging and opinion and experience sharing.

A first scenario that is commonly used concerns the presentation of pre-recorder lectures and the ability to study and comment these lectures by students and educators that are geographically distributed. For example Dr. Latchman, University of Florida uses slideshows of pre-recorder lectures with synchronized narration [8]. The students can use email, chat or forum in order to communicate with the educator, to oppose an opinion or discuss for a certain subject proposed by the educator and related to certain parts of the slideshow.

Fig. 1. Presentation of pre-recorded lectures with synchronized narration
[Source: http://www.clickandgovideo.ac.uk/]

Another way that we can combine streaming media with collaborative learning is to collaboratively annotate streaming videos. For example consider the scenario where a group of medical students use an AMCL system that supports video annotations to analyze and annotate a heart surgery video. At first the educator can create annotations (e.g. text, graphical or audio) representing tasks in order to enhance motivation and develop interaction among the students! In order to complete their tasks the medical students can use various tools supported by the AMCL system in order to communicate, express their opinion and interact with each other.

3 An Overview of the Existing Asynchronous Multimedia Collaborative Systems

AMCL systems combine the richness of multimedia representation and demonstrations of practice with the flexibility in the use of time for communication. The AMCL is a rather new education medium and philosophy, still unexplored. A well designed AMCL system supports functions specially designed for viewing and manipulating the audiovisual streaming material for the needs of performing collaboration learning activities and in addition the usual administration functions. Some of the functions that these systems support are:

- Support of three at least types of users (Student, Teacher and administrator)
- Conference management. This includes creation, deletion and modification of a conference by an authorized user usually the administrator or teacher.
- Playback of audio-visual content for the current conference.
- Annotating a certain frame of the audio-visual content. Annotations may be text, audio or drawings.
- Support of a "user portfolio" where the user can store important or personal messages.
- Advanced video and audio processing.
- User and message statistics such as number of posted messages, types of messages etc.

Various AMCL systems have already been developed and tested especially for learning environments. Some examples are stated below:

3.1 Project Pad

Northwestern University developed a project called "Project Pad" in order to build a web-based system for media annotation and collaboration for teaching and learning and scholarly applications. It consists of various tools such as the "Image Tool", the "Transcript Tool" and the "Video and Audio Tools". The "Video and Audio Tools" lets you attach comments to time segments of Flash FLV video and MP3 audio streams. The tools can be used by instructors and / or student teams to critique student-produced video and audio or to provide a way for students to analyze scientific, historic, or artistic recordings.

The tools feature a timeline that you can zoom in to mark detailed events or zoom out to annotate larger segments. Annotations are represented by markers that

Fig. 2. The "Video and Audio Tools"
[Source: http://dewey.at.northwestern.edu/ppad2/09road_map.html]

you can drag and re-size with the mouse. Attached text can include multiple fonts, font sizes, and styling.

3.2 VAnnotea

VAnnotea is a Collaborative Video Annotation tool that supports Collaborative indexing and annotation of audiovisual content over broadband networks [http://dewey.at.northwestern.edu/ppad2/09road_map.html].

The tool was developed by the School of Information technology and Electrical Engineering of the University of Queensland in Australia.

VAnnotea has a lot of features including browsing through existing online multimedia repositories using the embedded Internet Explorer Browser, viewing a wide variety of media formats such as MPEG-1, -2 and -4, WAV, MP3 and QTVR through embedded media players such as the Quicktime Player, Windows Media Player and Video Lan Client and annotating the media files by highlighting regions and attaching personal notes, questions, remarks, links and relationships to other resources, terms from ontologies or controlled vocabularies, ratings and local files such as images, or PDF documents.

Vannotea's flexible design and metadata architecture allows it to be used within many other application domains, including: Biology (Integrative Biology VRE), Oceanography and Marine biology etc.

3.3 XMAS

The MIT University has developed a video annotation system called XMAS in cooperation with Microsoft in order to support the study and comparison of Shakespeare texts, images and films. Learners and practitioners were able to

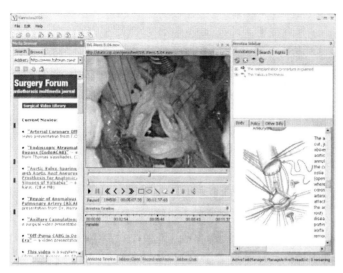

Fig. 3. VAnnotea - A collaborative Video Annotation tool
[Source: http://dewey.at.northwestern.edu/ppad2/09road_map.html]

watch video clips of theatrical plays, analyze them and participate in asynchronous discussion through discussion forums.

XMAS is currently optimized for use with commercially available DVDs as video source. XMAS allows users to rapidly define segments of film which can be replayed by clicking on automatically created links that can be saved in a list or dragged and dropped into discussion threads or online essays.

3.4 Video Traces

The Video Traces system is another system that enables users to use video in a collaborative way. The system allows audio annotations on specific video frames and in addition supports video processing functions like adjustable video speed, pause, rewind and fast forward.

The annotations are listed in a different window and can be sorted by title, author, date and time. The resulting product (video+ annotations) is called a "video Trace". A video trace can be further annotated for a variety of teaching and learning purposes or exchanged with other users.

The system was used in various educational scenarios like an undergraduate choreography class at the University of Washington.

3.5 iVAS

The iVAS system is a system that enables users to associate any video content on the Internet with annotations. The system was developed by Nagoya University in Japan and supports a lot of features such as text annotations, impression annotations, automatic evaluation method of annotation reliability, video simplification, and video-content-based community support.

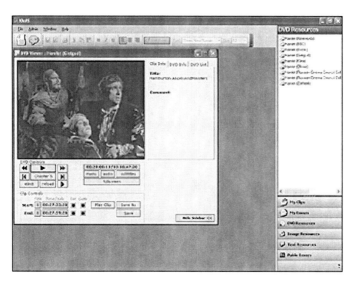

Fig. 4. The XMAS Video Annotation System
[Source: http://icampus.mit.edu/projects/xmas.shtml]

Fig. 5. The Video Traces annotation tool
[Source: http://depts.washington.edu/pettt/projects/videotraces.html]

In order to evaluate the iVAS system's usability and data collection, an experiment with 30 college students was performed. They used 5 minutes long video clips with various content such as news, drama, variety, and cooking program. The college students had to use the system to annotate the videos and then answer a questionnaire concerning the AMCL system.

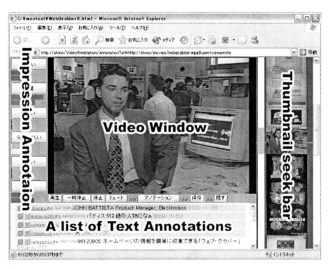

Fig. 6. The iVAs system
[Source: http://www.nagao.nuie.nagoya-u.ac.jp/ivas]

3.6 ISEE

ISEE is another AMCL system that is specifically designed for low bandwidth network users [9]. ISEE supports the usual video playback functions, as also and video annotation functions. A full version of ISEE contains a video player, an interactive chat room, a built-in web browser, and a story board.

When a user starts writing a note in the input box the system freezes the video and it continues when he press the submit button. The user can also apply time stamps that link the notes to video segments.

The system was tested by university students. The students had to comment their colleagues' presentations using the annotation and the playback functions.

3.7 KMI's Stadium

Another very interesting approach is the Stadium system of the UK Open University's KMI. The system is based on Webcasting technologies and allows the users to send short messages (similar to sms) during the video playback. Several companies used Stadium to train their employees [http://cnm.kmi.open.ac.uk/projects/stadium].

For example the Wytch Farm Bp Company located in Dorset England used Stadium to train their employees for security issues concerning oil pumping. Several company employees watched a pre-recorded audio-visual content in their offices located in different places in the world such as Bogotá, Houston, London and Aberdeen. At the end the employees were very pleased with the whole idea and found the system and content very interesting and helpful for their work.

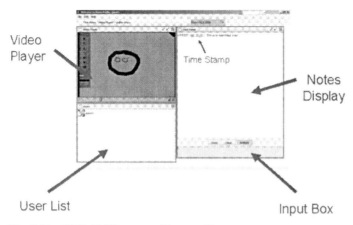

Fig. 7. The ISEE AMCL system [Source: 9]

4 Comparing the AMCL Systems

Although the presented systems have unique functions and are valuable for AMCL they support only some of the required functions of a well designed AMCL system. Some of them are specialized in video annotation and do not or partly support collaboration or administration functions such as threaded discussions and message management respectively while some others do not support video annotation at all. Some of the most significant limitations of most of the systems are stated below:

- Limited collaboration features
- Poor support for Audio, Text and Graphical Annotation features
- Streaming technologies are partly supported or not supported at all
- Limited playback and video analyzing features
- Multilanguage support is partly supported or not supported at all
- In most cases these systems support only Windows Platforms
- Luck of Vector Graphics support
- High hardware and software Requirements
- They support only some of the major video formats
- Require users with advanced computer skills
- Poor Help support

In Table 2 we can see a comparison between some of the current AMCL system and a desired AMCL system. The comparison is based on five feature categories: Video Analyzing, Annotation, general and miscellaneous features. Many of the desired features were derived through hypothetical usage scenarios of AMCL systems in various research areas. Two of these scenarios are described later on this chapter in order to derive important service and functional requirements.

The following use case diagram (Fig. 9 in Section 5.2) gives a graphical overview of the functionality needed by the desired AMCL system in terms of actors, their goals, and any dependencies between those use cases.

Table 2. Comparison of features between AMCL systems

Feature	Desired AMCL System	Project Pad	VAnnotea	XMAS	Video Traces
Video Analyzing					
Media Format	Most Important media types e.g. Avi,mpeg,flv,mp3,wav	FLV,Mp3,WAV	Mpeg 1-2-4, Mp3	DVD	Mpeg 1-2-4, AVI
Playback Functions	Play, Pause, Stop, Frame Seek, Jog wheel, seeking bar, Loop	Play, Pause	Play, Pause, Fast Forward, Rewind	Play, Pause, Stop, Fast Forward, Fast Rewind, Rewind	Play, Pause, Stop, Fast Forward, Fast Rewind, Rewind
Looping	Yes	No	No	No	No
Volume Functions	All Volume functions (Up, Down, Mute)	Volume slider	Volume slider	Volume slider	Volume slider / Mute
Segments	Yes	Yes	Yes	Yes	No
Skip (Jump) Segment	Yes	No	No	No	No
Zoom In / Zoom Out	Yes	No	No	No	No
Move Frame	Yes	No	No	No	No
Seeking Methods	Advanced seeking functions (Jog wheel, Seeking bar, Shortcuts etc.)	Seeking bar	Seeking bar	Seeking bar	Slow / Fast Motion
Layered Style timeline	Yes	Yes	Yes	No	No
Annotations					
Type	Text, Audio, Graph, URL, Metatags, File Attachment	Text	Text, Graph, URL, File Attachment, Metatags	Text, File Attachment (Image & Text)	Audio
Actions	Add, Delete, Edit, Goto	Add, Delete, Goto	Add, Delete, Goto	Add, Delete, Goto	Add, Goto
Load Annotations	Automatic / Manual (Import XML file)	Automatic	Automatic	Automatic	Automatic
Merge Annotations	Yes	No	No	No	No
Annotation Access Rights	Yes	No	Yes	No	No
Sorting by	Type, Author, Title, Time	?	?	?	title, author, date and time
Collaborative Annotation	Asynchronous	Asynchronous	Synchronous (Jabber) & Asynchronous	Asynchronous	Asynchronous
Output Format	Various (XML, TXT etc)	?	?	?	?
Graph. Annotations	Supported	Not Supported	Supported	Not Supported	Not Supported
Shapes	Free hand, line, circle, box, curve, triangle	Not Supported	Box, Circle, Line	Not Supported	Not Supported
Functions	Draw / Clear drawing / Edit Drawing/ Resize / Undo / Fill	Not Supported	Draw/Clear drawing / Undo/ Fill	Not Supported	Not Supported
Num of Colors	Multiple	Not Supported	?	Not Supported	Not Supported

Table 2. (*continued*)

	Supported	Not Supported	Not Supported	Not Supported	Not Supported
Audio Annotations					
Record Sound	Yes (From Input Device)	Not Supported	Not Supported	Not Supported	Yes (From Input Device)
Playback	Yes	Not Supported	Not Supported	Not Supported	Yes
Audio Format	Mp3 / WMA / WAV	Not Supported	Not Supported	Not Supported	WAV
General					
User Authentication	Yes	No	No	Yes	No
Bandwidth Detection	Yes	No	No	No	No
Customized Shortcuts	Yes	No	No	No	No
Vector Graphics	Yes	Yes	No	No	No
Streaming Technology	Progressive Download, Real Streaming	Progressive Download	Real Streaming	Not Supported	Not Supported
Upload File	Yes	No	Yes	Yes	No
Attach URL	Yes	No	Yes	Yes	No
Attach File	Yes	No	Yes	Yes	No
Portfolio	Yes	No	No	No	No
Multilanguage Support	Yes	No	No	No	No
Export to Learning Object	Yes	No	No	No	Video Traces (Media + Annotation)
Export Log File for Data Analysis	Yes	No	Yes	No	Yes
Misc					
Application Type	Multiplatform System	Windows 95,98,XP,MacOSX,Linux	Microsoft Windows XP	Microsoft Windows Xp	Windows 95,98,XP
Requirements	Minimum	Flash 7 plug in	Microsoft .NET Framework Version 2.0, Windows Media Player 10, QuickTime 7.1	512MB Ram, DVD Rom, DSL (Min 384 Kbps)	Windows Media Player Classic

5 Learning Scenarios for Extracting Requirements for AMCL Systems

In order to design and develop an AMCL system that does not suffer from the limitations described before we can consider the following authentic scenarios. These scenarios could help us derive all the necessary functionality and service requirements of a well designed AMCL system. The first scenario concerns the "Microteaching" technique. According to [10] Microteaching is a simulation technique which enables us to create or modify a desirable or undesirable type of the instructor's teaching behavior. The main goal of Microteaching is to give instructors confidence, support, and feedback by letting them try out among friends and colleagues a short slice of what they plan to do with their students.

Usually for the needs of microteaching, teachers give lectures which are videotaped, reproduced and discussed by an evaluation team which consist the instructors and a teaching consultant. This traditional type of Microteaching can be transformed into a new form if we use AMCL systems and audiovisual content to create a new flexible environment for instructors distance training.

5.1 Microteaching for Civil Engineers

In this scenario we will demonstrate a usage scenario of an AMCL system which can be used to train civil engineers about the basic features of AutoCAD. For the needs of the instructional scenario, we assume that the civil engineers are organized in groups of four. Firstly the educator uses **a GUI** (Web Interface) in order to **search for a Streaming server** responsible to store and stream educational video clips and a **File Server** responsible to store other files such as text documents, images etc. This is possible through a **search engine** for distributed services. He also uses an **upload service** to upload a pre-recorder video clip in streaming format containing a tutorial for AutoCAD's basic features and a text document (containing other important information related to AutoCAD functionality) on the **Streaming Server** and the **File Server** respectively. He requests and uses a **video processing service** in order to **convert** various formats such as (avi, mpeg, etc) into streaming format.

In the next step he/she uses the AMCL system to **login** as "Educator" through an **authentication service** in order to create (**write**) three new annotation points that represent three tasks (one for each group). The AMCL system creates automatically an annotation file that contains such as "Type of annotation", "Time Position", "Description" etc and was also uploaded by the educator on the file server.

Each group uses the AMCL system to log in as "Learner" enabling them to have certain **File Access Rights**. They also request the **list of available video clips** in order to select and watch the uploaded video clip and **read the annotation points** containing text and graphical information about their task. They also use some advanced processing features to reprocess and analyze the video clip, to mark, comment and pose questions on selected frames of the video clip. In order to get more information related to AutoCAD's features, they use the text documentation uploaded by the educator.

In the last step the groups submit a text file containing a report of their conclusions and comments about the procedure they followed to complete their tasks. At the end the educator and the practitioners discuss and analyze each group's report and suggest other possible solutions. The educator has also posed questions concerning the system and asked practitioners to give feedback and any suggestions they may have.

5.2 Collaborative Basketball Training

In this scenario we will demonstrate the usage of an AMCL system in order that a coach, the technical manager and the players of a basketball team to collaboratively analyze and annotate various videos from previous basketball games. They want to watch this video material in order to decide about the system (tactic) that the team will follow during the next matches, to improve their playing and to reduce turnovers.

To begin with the administrator of the system uses an **upload service** to upload the pre-recorder video clips and other important documents related with basketball tactics on a **Streaming Server** and the **File Server** respectively. He also uses a **video processing service** in order to prepare the video for streaming!

In the next step the coach and the technical manager login as "Educator" through an **authentication service** in order to create new annotation points that represent new tasks (A question, a comment etc) in order to motivate the players to start an asynchronous communication, exchange of information and experience and to start analyzing the important scenes during the games!

The players use the AMCL system to log in as "Learner" enabling them to have certain **File Access Rights**. They also request the **list of available video clips** in order to select and watch the uploaded video clips and **read the annotation points.** They also search frame by frame the videos using advanced video editing features in order to find important scenes such successful gaming, defense systems and turnovers. After that they can use the video playback features in order to annotate the selected scenes. The AMCL system enables users to create various types of annotations such as text, audio and Graphics. Graphical annotations can be created using special tools for drawing such as free hand drawing, drawing shapes (boxes, circles, curves, and lines), and clear drawing in order to successfully annotate an important scene. For example they can annotate a scene where the famous "pick & roll" tactic is successfully.

Finally they can choose to create Text or Audio Annotations in order to comment scenes and express thoughts about their team. After the end of the analyzing and annotation procedure the whole team will watch the annotated video and the technical team will explain the systems that the team must follow in the next games.

In both aforementioned scenarios a streaming server containing the streaming content (video and audio), an application server where to store the AMCL system and an annotation server to store annotations are needed. The AMCL system interacts with a database for the user authentication process and uses the streaming technologies to deliver the audiovisual content. Users communicate and annotate

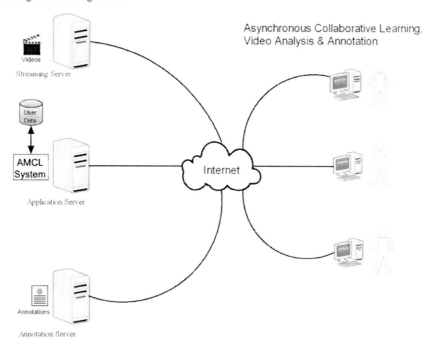

Fig. 8. General architecture of an AMCL system based on the service and functionality requirements

the audiovisual content in a collaborative way and so they enrich audiovisual content with semantic information and enhance the learning activity. Figure 8 illustrates the general architecture of an AMCL system.

The Functional requirements from an AMCL system, as extracted from the scenarios, are the following:

- Support of three at least types of users (Student, Teacher and administrator)
- Conference management. This includes creation, deletion and modification of a conference by an authorized user usually the administrator or teacher.
- Playback of audio-visual streaming content for the current conference. This also includes Advanced playback functions such as play, stop, frame by frame search, loop.
- Annotating a certain frame of the audio-visual content. Annotations may be text, audio or drawings.
- Support of a "user portfolio" where the user can store important or personal messages.

Figure 9 shows these requirements in a more illustrative version by using the UML use case diagrammatic notation.

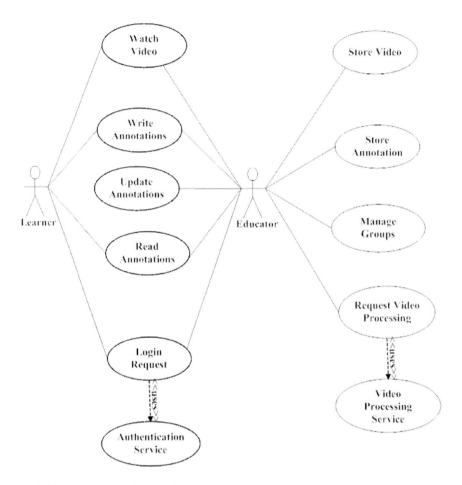

Fig. 9. Use case diagram for both "Learner" and "Educator"

6 New Development Trends in the Domain of AMCL Systems

The previous analysis of the current AMCL system and the two learning scenarios showed us that there is a need for new functionality & architectural decisions. AMCL systems must be developed in a different way, supporting of course a wider range of functionality, collaboration and video annotations. AMCL systems must be developed with state of the art technologies, supporting latest streaming technologies and taking advantage the high bandwidth on internet connections that we enjoy nowadays. They should support asynchronous discussion and exchange of viewpoints about vivid representations of practice in the form of video or audio clips.

Furthermore, they could support textual, graphical or audio annotations on specific frames of the presented audiovisual material. Powerful annotation mechanisms are very important to be integrated into the AMCL systems because they will enable learners, teachers and other practitioners to explain issues, share opinions, mark important frames and pose questions, over time and anywhere in space.

Other specialized features which could facilitate the collaboration among students could be offered by AMCL systems such as:

- Support of a portfolio where the student can store, view or delete video clips, annotations on the video clips as well as specific comments/messages made by users during the online discussion.
- Presentation of statistics where the student can see how many messages have been posted by how many users as well as the number of the different types of messages (text, videos, audios, still images).

7 Conclusions

In this chapter we described the importance of using streaming media for collaborative learning purposes. We also stated various research and development achievements in this domain focusing on AMCL systems. Nowadays new systems are starting to appear which try to eliminate the usability problems of the existing systems as well as offer new set of features that will enable students collaborate more easily. This domain though in its infancy is fast growing. We are living an era where videos are easy to be created, found and shared among peers. Sites like YouTube are proofs of this statement.

Educators and educational technologies have the challenge of creating systems like AMCL that will make students become active consumers of digital audiovisual learning content.

References

1. Severin, W.J.: Another look at cue summation. Audio Visual Communications Review 15, 233–245 (1967b)
2. Hartsell, T., Yuen, S.: Video streaming in online learning (2006)
3. Bowie, M.M.: Instructional film research and the learner. Paper presented at the Annual Convention of the Association for Educational Communications and Technology, Las Vegas, NV (ERIC Document Reproduction Service No. ED 267 757) (1986)
4. Salomon, G.: Interaction of Media, Cognition, and Learning, p. 16. Lawrence Erlbaum Associates, Hillsdale (1994)
5. Wetzel, C.D., Radtke, P.H., Stern, H.W.: Instructional Effectiveness of Video Media. Lawrence Erlbaum, Hillsdale (1994)
6. Goodyear, P., Steeples, C.: Creating shareable representations of practice. Advance Learning Technology Journal (ALT-J) 6(3), 16–23 (1998)
7. Young, C., Asensio, M.: Looking through Three 'I's: the Pedagogic Use of Streaming Video. In: Banks, S., Goodyear, P., Hodgson, V., McConnell, D. (eds.) Networked Learning 2002, Proceedings of the Third International Conference, Sheffield, March 2002, pp. 628–635 (2002)
8. Thornhill, S., Asensio, M., Young, C.: Video streaming: A guide for educational development (2002), http://www.clickandgovideo.ac.uk/

9. Fu, X.: Marchionini, & Pattee, Video Annotation in a Learning Environment (2003)
10. Allen, D.W., Ryan, K.A.: Microteaching. Addison-Wesley, Reading (1969)

URLs

11. VAnnotea: Collaborative indexing, annotation and discussion of audiovisual content over high bandwidth networks, http://www.itee.uq.edu.au/eresearch/projects/vannotea/index.html
12. Project, PAD: a web-based system for media annotation and collaboration for teaching and learning and scholarly applications, http://dewey.at.northwestern.edu/ppad2/09road_map.html
13. XMAS: A Video Annotation tool to support the study and comparison of Shakespeare texts images and films, http://icampus.mit.edu/projects/xmas.shtml
14. Video Traces: Rich Media Annotations for Teaching and Learning, http://depts.washington.edu/pettt/projects/videotraces.html
15. Heart Surgery Forum: An online video library, http://www.hsforum.com/svl
16. Yamamoto, D., Nagao, K.: iVAS: Web-based Video Annotation System and its Applications. In: proceeding of ISWC 2004 (2004), http://www.nagao.nuie.nagoya-u.ac.jp/ivas
17. KMi Stadium, http://cnm.kmi.open.ac.uk/projects/stadium
18. Using video in Education, http://www.libraryvideo.com/articles/article13.asp

'MULTIPLES': A Challenging Learning Framework for the Generation of Multiple Perspectives within e-Collaboration Settings

Maria Kordaki

Department of Computer Engineering and Informatics, Patras University, 26500, Rion, Patras, Greece

Abstract. This chapter proposes a learning framework - the MULTIPLES framework (Multiple: Learning Tools, Interactions, Perspectives, Learning Experiences and Solution Strategies) - that can be used in e-collaboration settings to encourage the development of multiple perspectives for each individual student regarding the learning concepts in focus. This framework has been designed within the context of social and constructivist learning theories, acknowledging the role of asking learners, working in both groups and individually, to face appropriately designed learning tasks by using multiple learning tools and representation systems while at the same time performing various interactions in order to produce multiple solution strategies *'in as many ways as possible'*. To this end, a case study is reported that illuminates the role played by MULTIPLES in the enhancement of each individual student's views by generating different solution strategies to the tasks at hand, while at the same time expressing their inter- and intra-individual differences. Specifically, 25 secondary level education students participated in a learning experiment consisting of two tasks asking for the construction of multiple solution strategies regarding the concept of area measurement by exploiting the plethora of tools and representation systems provided by the well-known educational software Cabri-Geometry II (Laborde, 1990). The analysis of the data shows that all these students truly exploited their collaborative experience during the former task and constructed various solution strategies when individually facing the second task, at the same time developing multiple perspectives of the aforementioned learning concepts.

1 Introduction

Recent advances in Internet and Web-based technologies have provided educators with great opportunities to reconceptualize learning by extending both its boundaries and pedagogies in terms of: learners' communication capabilities, diversity of learning interactions, releasing learning from time and space, contributing towards equal learning opportunities, and providing schools with opportunities to exchange learning experiences, so they can acquire new perspectives and broaden their social horizons [13, 28, 4]. Most importantly, the Internet has been widely recognized as a medium that allows for the design of learning environments based on modern constructivist and social theories regarding teaching and learning [12, 17, 9]. Distributed and situated cognition views of knowledge construction can also be considered to interpret learning events within the framework of networking technologies [33, 36, 7].

Within the context of these modern theories, enhancing learner perspectives in terms of the learning concepts in focus, while at the same time acknowledging their inter- and intra-individual learning differences, is essential [27]. To motivate learners to develop such diverse learning perspectives, the role of appropriately-designed learning activities is crucial. To this end, a collaborative learning framework - the MULTIPLES framework - has been formed. This framework is situated firmly within the context of modern constructivist and social theories of learning, where the learning process is viewed as an active, subjective and constructive activity within authentic contexts rich in computer learning tools and communication interactions [40, 31, 18]. Specifically, within the MULTIPLES framework, learning is emphasized in a context supporting: a) the learning of essential aspects of each learning subject [19], b) open multiple solution-based authentic learning activities [16, 20, 38, 22], c) the expression of learners' inter- and intra-individual differences [27, 24], d) active learning using multiple computer learning tools and representation systems [21], e) the generation of multiple solution strategies to the tasks at hand by both group and individual learners, f) multiple ways of interaction and communication [13], g) appropriate intrinsic and extrinsic feedback for self-correction [32], and h) multiple ways of assessment [18]. It is worth mentioning that collaborative learning frameworks supporting the performance of multiple solution-based tasks asking to be solved '*in as many ways as possible*', while at the same time exploiting the availability of multiple learning tools and representation systems, have yet not been reported.

The role of the proposed framework was investigated through its use in the design of a learning experiment – using real students - for the learning of concepts related to conservation of area and its measurement by secondary level education students. These concepts are viewed as being essential to students' mathematical learning [14]. This chapter is organized as follows: in the next section, the background of MULTIPLES is presented and, subsequently, its architecture is demonstrated. Then, the context of the aforesaid specific learning experiment is reported while the emerging empirical results are discussed in relation to the specifications of the proposed architecture of MULTIPLES. The chapter ends with the conclusions.

2 Background

E-Learning can be defined as an open and distributed environment that provides learning tools, enabled by Internet and Web-based technologies, to facilitate knowledge building through diverse, new and meaningful learning actions and interactions such as learner-learner, learner-diverse groups, learner-various type of resources, learner-instructor, learner-expert and learner-cognitive tools. In fact, within the Web-based context, learning can be considered as a function of a plethora of interactions with others and with various tools to face appropriately-designed tasks. In this context, learning can also be reconceptualized as a social and distributed process over time and place, both synchronously and asynchronously, where the concepts of learning in groups, sharing information, meaning negotiation and co-construction of knowledge are dominant, and the concept of distance is not significant. However, the importance and necessity of linking appropriate learning theories in the design of e-learning environments has been acknowledged by many researchers, despite the aforementioned challenging teaching and learning capabilities recently facilitated in such environments [6, 1].

Understanding the coordination among individuals and artifacts in a system or community is a main principle of distributed or situated cognition in terms of knowledge construction [33, 36]. This view suggests that cognition is not an event that takes place inside one's head in isolation, but can be looked at as a distributed phenomenon that is more global in nature—one that goes beyond the boundaries of a person to include environment, artifacts, social interactions, and culture [35]. Subsequently, it is acknowledged that individuals learn from activity and the tools supporting it to extend their cognitive capabilities [30]. The situated cognition view is consistent with the epistemological assumptions of constructivism, and social views of learning which stipulate that meaning is a function of how the individual creates meaning from his or her experiences and interactions within authentic contexts [16, 18]. In fact, promoting authentic learning activities is a central aspect of constructivist learning that is also viewed as an active, subjective and constructive process. On the whole, it is worth noting that the context or activity that frames knowledge construction is of equal importance to the learner as the knowledge itself [6].

Authentic activities engage learners in realistic and meaningful tasks that are relevant to each learner's interests and goals, motivating him to be actively involved in their learning [41, 30]. Authentic learning tasks emphasize the encouragement of learners to explore different solution strategies while simultaneously forming various hypotheses and observing their effects [3, p. 135]. Hypothesis generation and exploration goes hand by hand with the acquisition of problem-solving and decision-making skills potentially essential in a learner's personal and professional life [6]. Main characteristics of the problem-solving approach also involve students working collaboratively in small groups, analysing and brainstorming ideas that could lead to a solution to a problem [10]. However, appropriately-designed learning activities should also be complex enough to require the cooperation of all members in order to work toward a solution, while also being open-ended and containing the content objectives of the course.

Social views of learning also suggest that contexts rich in cognitive tools and social interactions are constantly challenging the learner's understanding, resulting in new meanings [31]. Learners are provided with the opportunity to transform their experience of performing authentic learning activities within collective socio-cultural settings into new knowledge [34]. Various studies have also reported the positive effects of collaboration within Web-based learning settings in different fields and disciplines [8, 29]. In collaboration and social negotiation, the goal is to collaborate to face the given learning activities and to share different viewpoints and ideas [5].

In fact, communication within the context of Web-based settings can encourage learners to acquire a rich and robust knowledge base by developing multiple perspectives regarding the learning concepts in focus through the promotion, articulation and negotiation of - and reflection on - different and contrasting views. Novices can also develop their competences by exploiting their collaboration experience within the aforementioned settings. In fact, when students are encouraged to articulate their knowledge to one another, they can share multiple perspectives and make generalizations that can be applicable in a number of different contexts [3]. Promoting students' reflective thinking involves asking them to review, analyze and concentrate on their experience in order to extract the most important and useful points of view on both the learning concept in focus and the process of learning itself [3]. It is also worth noting that, when students are working with peers instead of alone, anxiety and uncertainty are reduced as they collectively find a way to face the given tasks and they find these tasks interesting and satisfying [13].

Promoting the development of multiple perspectives during collaborative learning settings is a learning approach that provides opportunities for the construction of flexible and meaningful knowledge structures [5, p. 178], at the same time encouraging the expression of learners' inter- and intra-individual differences [24]. Here, it is worth noting that student inequality in learning and achievement at school has been linked to their inter-individual learning differences [37, 39]. Most learner difficulties are found in the gap between their intuitive knowledge and the knowledge they need to express themselves in the proposed representation systems [15]. For example, prepositional, symbolic and abstract representation systems prevent some learners (usually beginners) from expressing their knowledge, the same systems being intended for use by advanced learners. Being provided with opportunities to express different perspectives on the task at hand, learners are also given opportunities to express their own knowledge, including their mistakes, as well as to master more than one learning approach. To this end, the use of multpile representation systems is acknowledged as crucial in encouraging the expression of learners' inter- and intra-individual variety regarding the subject to be learned [15, 23].

Being exposed to multiple points of view, students also have the chance to develop broad views regarding the learning concepts in focus [22]. Essentially, emphasizing multiple perspectives involves not only presenting information in various ways but encouraging learners to use multiple learning tools and representation systems to construct their own multiple solution strategies for the tasks at hand and also document their own explanations. With the promotion of multiple perspectives, learners also have the chance to become aware that there are multiple approaches to an issue, which is the case in real life situations. In addition, learners have the chance to explore each perspective to seek a meaningful resolution to the issue at hand, constructing new meaning within the context of their own experiences and knowledge [6].

3 The General Architecture of MULTIPLES

In this section, architecture of a typical representative of MULTIPLES is presented. The construction of this architecture was based on modern constructivist and social theories mentioned in the previous section, also exploiting the advantages provided by the Computer Based/Internet Based technologies. On the grounds of these learning theories, a number of specifications were formulated and used in the construction of the said architecture, which consists of four main parts: (A) learning activities, (B), learner activity space (C), learner communication and (D) learner assessment. The description of the design of these parts is presented below:

Part A. *Learning activities.* This part of the MULTIPLES framework includes basic specifications related to the design of constructivist learning activities suitable in a computer context. In terms of design specifications, these activities have to:
a) focus on both the fundamental aspects of the learning subject and the specific learning points where the learners illuminate difficulties, b) be drawn from the context of the learners' everyday life, c) encourage problem-solving skills (that is, stimulate learners' higher mental functions in terms of analytical and synthetical thinking skills as well as critical thinking), d) promote collaboration and social negotiation, e) be solved in multiple ways using different aspects of the subject matter, as well as various representation systems, and also exploit the different kinds of knowledge that students possess, such as their previous knowledge, school-knowledge, intuitive knowledge, real life knowledge, visual knowledge, etc, f) demand each group construct solutions 'in as many ways as possible' by exploiting the provision of multiple tools and representation systems as well as the diversity of students' knowledge, g) demand each individual student perform similar tasks to those performed while working in groups, also 'in as many ways as possible' (in this way, each student would be able to exploit the advantages of his/her participation within the groupwork), h) allow students control over their learning (i.e. the activities can be solved using representation systems that provide appropriate feedback, such as intrinsic visual and/or explicit numerical), with learners being able to reflect on the feedback of their actions and then have the opportunity to correct their solution strategies), i) encourage learners to experiment by handling primary sources of data, at the same time acquiring hands-on experience, and j) not demand from learners extra complicated knowledge from other disciplines.

Part B. *Learner activity space.* This part of MULTIPLES includes specifications for the design of a *'learner activity space'*, that is, a virtual place where the learners can actively construct their own knowledge by performing the selected group of learning activities, using various cognitive tools. These tools have to be closely related to the specific learning subject and also appropriate to help learners perform the most representative and essential activities for the learning of this subject. The main aim of the integration of various cognitive tools within the 'activity space' is to provide learners with opportunities to: a) perform different solution strategies to the selected learning activities, thereby expressing their inter- and intra-individual learning differences, b) perform the same solution strategy using different representation systems, c) solve various activities for the learning of each specific learning subject, d) overcome basic difficulties regarding each specific learning subject, e) choose from among these tools those most appropriate for the expression of their knowledge, f) express different kinds of knowledge they possess, thereby developing a broad view of the concepts in focus.

Part C. *Learner communication.* Students have to be provided with various tools in order for them to be able to perform various communicative interactions, such as learner-learner, learner-teacher, learner-experts, learner-diverse groups, learner-whole class interactions. These interactions can be also performed synchronously and asynchronously. Learners can be encouraged to articulate their opinions and to reflect on both their experience and that presented by their colleagues. Learners can also be encouraged to share and negotiate their knowledge with their classmates. Small and whole class discussions can also be encouraged. To assist learners in performing all these communications, the learning framework can support a large number of synchronous and asynchronous conferences as well as publication of group/individual work and, of course, the use of e-mail.

Part D. *Learner assessment.* This part of the design of MULTIPLES includes specifications in relation to the formation of assessment activities. From the constructivist perspective, assessment can become a valuable tool for learning, the emphasis being placed on both group and each individual's learning processes and not exclusively on her/his learning outcomes. Multiple solution-based activities to be performed 'in as many ways as possible' can be used as a basic structural element of learner assessment. Thus, a combination of methods can be used, such as: a) self-assessment by reflecting on intrinsic and/or extrinsic feedback appropriately provided by the electronic learning environment in use, b) peer-to-peer-assessment by explaining, articulating and negotiating the solution strategies proposed by each individual student in front of both their groups and the whole class, and c) assessment by the teacher in terms of acknowledging the variety, the kind and the number of solution strategies constructed by both the groups and each individual student. This procedure could be realized by providing learners with opportunities to add electronic portfolios, posting their work over an extended period.

4 Using MULTIPLES with Real Students: A Case Study

Aim and methodology: The aim of this experiment is to investigate, through a comparative study, the role of the MULTIPLES learning framework in the development of each individual student's knowledge concerning the mathematical notion of area. To this end, students were provided with the opportunity to interact within the rich context of tools offered by the well-known educational software Cabri-Geometry II to face 'in as many ways as possible' two similar multiple solution-based problems in both a collaborative learning setting (1st task) and a setting where the emphasis is put on the actions of each individual student (2nd task). Such a comparative study has not yet been reported. In terms of methodology, this research is a qualitative study [2] focusing on the variety of solution strategies realized by students working with Cabri tools.

Cabri Geometry II [26]: Within the context of this software, constructivist mathematical learning settings can be supported. In fact, Cabri is highly capable of facilitating the design of learning activities that encourage learners to take an investigative perspective, express their inter-individual and intra-individual learning differences, make self-corrections, and formulate and verify conjectures [25]. In addition, authentic, meaningful, real life learning activities can be integrated within the context of this software. In particular, Cabri provides students with potential opportunities in terms of: i) *A rich set of tools* to perform diverse geometrical constructions according to various concepts in Euclidean Geometry, *ii) Tools to construct a variety of representations, both numerical and visual*, such as geometrical figures, tables, equations, graphs and calculations. These representations are of different cognitive transparency; consequently, students can select the most appropriate to express their knowledge, iii) *Linking representations*, by exploiting the interconnection of the different representation modes provided, iv) *Dynamic, direct manipulation* of geometrical constructions, by using the 'drag mode' operation, enhancing their knowledge about the issue at hand by dynamically exploring the invariance of their constructions, e) *The possibility of collecting large amounts of numerical data*. These data can be used by the students to form and verify conjectures regarding the geometrical concepts in focus, f) *Interactivity and multiple types of feedback* (intrinsic visual and extrinsic numerical), providing learners with opportunities to form and verify conjectures as well as self-correct their constructions, g) *Presenting information to the students in various forms*, h) *capturing the history of student actions* to provide teachers and researchers with a valuable amount of data for further studies, and i) *Extension*. Certain operations could be added as buttons on the Cabri interface following the formation of specific macros.

The learning experiment: The learning experiment took place in a typical, provincial, state secondary school in Patras, Greece [25]. Three complete classes of students participated in this experiment, consisting of: eight 1st grade students (13 years old), nine 2nd grade students (14 years old) and eight 3rd grade students (15 years old). These students were asked to perform two tasks using the Cabri II tools. The duration of tasks was commensurate with student needs; each student took about two hours to complete each task. Data resources include the electronic files of students' visual geometrical constructions, the video recordings of all interactions performed and the field notes of the researcher.

The tasks: Two tasks were assigned, the first being: a) to 'construct pairs of equivalent triangles, in as many ways as possible' using Cabri tools, b) to 'justify your solution strategy' and c) to 'explain what you think about the relation of the area and perimeter of these triangles'. Each student provided explanations and articulations of their solution strategies to her/his group as well as to the whole class. Whenever the students appeared to be on the point of giving up, the researcher would intervene, involving them in the task and encouraging them to continue by asking: 'try another way of constructing another pair of triangles with equal areas. You can use other tools and the different kinds of knowledge you possess'. Students worked in groups of three to perform this task, so as to exploit the advantages of cooperation. The learning aims of this task were to enable students to: i) advance smoothly from the notion of congruent triangles to the notion of equivalent triangles, beginning with the expression of their previous knowledge of congruent triangles and then enhancing their knowledge by exploring the equivalence in triangles using the 'drag mode' operation, ii) distinguish the concepts of area and perimeter in triangles by studying these concepts in relation to each other, and iii) link different kinds of knowledge about the concept of area through using the diversity of the tools provided, at the same time developing a broad view of this concept.

In the second task, students were asked to 'construct a triangle and to perform any possible sequence of modifications to produce other triangles equivalent to the original'. Specifically, students were asked to: a) 'construct an original triangle, then modify it into another equivalent triangle, using the Cabri tools', b) 'justify your solution strategy' and c) 'consider the produced triangle as the original triangle and repeat (a) and (b) as many times as you can, using different ways of modification'. The researcher intervened by encouraging the students to continue, as before. Students worked individually while performing the second task. It was decided to investigate how each individual student had perceived the learning experience of the first task after participating in the aforementioned group activity. The additional learning aims of this task were to enable each individual student to: i) construct individual approaches to the concepts of area and perimeter in triangles, ii) integrate different pieces of knowledge they possessed regarding the concept of area, iii) develop a broad view of the concept of area and its invariance by constructing a class of triangles equivalent to an original triangle through a sequential process of modification.

Communication: Multiple ways of communication were encouraged, such as: learner-learner during team work, learner-whole class during whole class discussions and learner-teacher during the whole experiment. Students also interacted with the tools provided by Cabri and received feedback in multiple ways. The whole experiment took place in a face-to-face communication setting, so as to focus on the investigation of the role of collaboration using the previously mentioned type of tasks within contexts rich in cognitive tools, such as Cabri.

5 Results

Group and individual solution strategies used by the students to face both tasks were classified into ten categories. Some of these strategies lead to the construction of congruent triangles (those presented in Table 1) while some others lead to the construction of equivalent not exclusively congruent triangles (those presented in Table 2). A further description of these strategies in terms of their value in mathematics education is presented in [25].

At first glance at the results presented in both Tables indicates that students exploited the presence of the plethora of the tools provided by Cabri as well as their collaboration within both small groups and the whole class, and constructed different solution strategies to the tasks at hand, at the same time expressing different kind of knowledge as well as their inter-individual differences and forming multiple perspectives.

Table 1. Categories of group and individual solution strategies to construct congruent triangles in the context of Cabri-Geometry II

Categories of strategies performed by the students to construct congruent triangles in the context of Cabri-Geometry II
Strategies were based on…
S1: students' visual perception and automatic control by measurement tools. *Tools used:* segment, automatic measurements of both area and perimeter.
S2: the preservation of the lengths or both; lengths and angles of the original triangle. *Tools used:* segment, automatic measurements of length, angle, area and perimeter.
S3: basic geometrical transformations such as: Translation (Strategy S3a), Reflection about an axis (Strategy S3b), Symmetry (Strategy S3c) and Rotation (Strategy S3d). *Tools used:* triangle, symmetry, reflection, rotate, translate as well as automatic measurements of both area and perimeter.
S4: the splitting of polygons: eg. Splitting: • an isosceles triangle using a perpendicular bisector (strategy S4a), • a rectangle & a square into two equivalent triangles by using one of its diagonals (strategy S4b), • a regular polygon into a number of equivalent triangles by using all its diagonals (strategy S4c), • a parallelogram into two equivalent triangles by using one of its diagonals (strategy S4d) *Tools used:* triangle, perpendicular line, median, polygon, regular polygon, parallel lines, circle as well as automatic measurements of both area and perimeter.
S5: the formation of specific geometrical constructions producing pairs of congruent triangles. *Tools used:* parallel and perpendicular lines, lines, segments, triangle, the 'drag mode' operation, rotation about an angle and around a point, automatic measurement of both area and perimeter, automatic tabulation of numerical data.

Specifically, students expressed intuitive approaches to area (strategy S1), spatial approaches in terms of area measurement using spatial units (strategy G4), formal approaches in terms of using the area formulae (strategies G2, G3, G5) and dynamic approaches (strategies G1, G2 and G5b) where they investigated the possibility of the existence of shapes with equal areas and different forms. Students also expressed their mistakes in terms of shapes of equal perimeters having equal areas (strategy S2). Finally, students related area to different kinds of knowledge they possessed, such as basic area transformations (strategies included in category S3), polygons (strategies fall into category S4) and specific geometrical constructions formed by using arbitrary, parallel and perpendicular lines (strategy S5). In performing all the aforementioned strategies, students also exploited the possibility of the direct manipulation of their constructions by using hands-on experience and formed/verified hypotheses and conjectures by reflecting on both the visual and numerical data automatically produced by the system. Students also exploited the multiple ways of feedback (intrinsic and numerical) provided by Cabri to verify their solution strategies at both group and individual levels. The feedback given by their teacher was also essential.

Table 2. Categories of group and individual solution strategies to construct equivalent not exclusively congruent triangles in the context of Cabri-Geometry II

Categories of strategies performed by the students to construct equivalent not exclusively congruent triangles in the context of Cabri-Geometry II
Strategies based on:
G1: the investigation of the possibility of the existence of equivalent triangles using the 'drag' mode in combination with automatic area measurement: *Tools used:* triangle, the 'drag mode' operation, as well as automatic measurements of both area and perimeter.
G2: the conservation of both; the length of the base and its distance from the opposite vertex in a triangle. *Tools used:* parallel lines, segment, point, triangle, the 'drag mode' operation, automatic measurement of both; area and perimeter, automatic tabulation of numerical data.
G3: the splitting of a triangle into two equivalent triangles using a median. *Tools used:* triangle, segment, median, automatic measurement of both area and perimeter.
G4: measuring areas using area-units: *Tools used:* triangle, square grid, segment, the 'drag mode' operation, median, automatic measurement of both area and perimeter.
G5: using area formulae. Constructing an original triangle ABC and trying to construct another triangle equivalent to the original by conserving the product of the lengths of its base and its respective altitude, eg. in right-angled triangles (Strategy G5a), in arbitrary triangles (Strategy G5b) or in an arbitrary triangle at the same time sliding the altitude on the base of the triangle. *Tools used:* triangle, perpendicular line, segment, measurement of length, calculation, , the 'drag mode' operation, median, automatic measurement of both; area and perimeter, measurement transfer, automatic measurement of both area and perimeter.

In the following section, the specific strategies constructed by both each group and each individual student are presented.

5.1 Group and Individual Multiple Solution Strategies to the Tasks at Hand Using the Tools Provided by Cabri-Geometry II

Group and individual multiple solution strategies to face both tasks are presented in Table 3.

Table 3. Group and individual multiple solution strategies performed within Cabri-Geometry II

Grades	Group	Strategies performed by each group	Total Strat.	Students	Total Strat	Strategies performed by each student
1st	A1	S3a, G1, G4, G2, G2, S4a, S4b, S3b, S3c, S5, G3	11	P1	6	S3a, G1, S3b, S2, G2
				P2	6	G1, S3a, S4a, G2
				P3	6	S3c, S1, S3d, G3, G2
	A2	S1, S4c, G2	3	P4	3	S1, S4c, G2
				P5	4	S1, G4, S3a, G2
	A3	G1, G2, S3b, S4c	4	P6	3	S3a, G1, S3b

'MULTIPLES': A Challenging Learning Framework

Table 3. (*continued*)

				P7	7	S1, S3b, S4c, G1, G2 S3d
				P8	5	S3b, G4, S3d, G1, G2
2nd	B1	S3a, G1, G2, G4, G5a, S4d, G3	7	P9	6	G4, S3a, G3, G5a, G2
				P10	7	S3b, G3, S3c, G1, G5a, G2
				P11	6	G5, S4d, G2, S3
	B2	S1, G1, S3c	3	P12	5	G4, S3c, S4
				P13	3	S4d, S3b, G1
				P14	6	S3a, G1, S3c, G2, G5a
	B3	G5b, S2, G1, G4, S3b, S3d, G3	7	P15	4	G1, S3b, S2
				P16	4	S3b, G1, S3d
				P17	9	G4, S2, G2, G5b, S3b, S4c
3rd	C1	S1, S3d, G1, G4	4	P18	3	S3a, G1
				P19	3	G4, G2, S3d
				P20	5	S1, G1, S3b
	C2	S3a, G3, S3b, S3c, G1, G2	6	P21	4	G1, S3a
				P22	4	G1, S3b
				P23	4	S3c, G2
	C3	S1, G1	2	P24	7	S3a, G1, G5b, S4b, G3, S3c
				P25	3	S3b, S1, G1

In this Table, the capital letters A, B, C are used to represent the 1st, 2nd and 3rd grades, correspondingly. These letters are used in combination with the numbers 1, 2, 3, to represent the number of groups in each respective grade that performed the specific strategy. Capital letters Pi (i=1…25) are also assigned to represent each individual student participating in this experiment. As is clearly demonstrated in this Table (3rd column), each group exploited the presence of the various tools provided by Cabri and the collaboration during the 1st task and expressed their inter-individual differences to perform multiple solution strategies for this task, while at the same time constructing multiple perspectives of the concepts in focus. It is also clearly shown (see the last column of Table 3) that each student exploited their collaborative experience during their work in small groups and within the whole class to perform multiple solution strategies to the 2nd task, at the same time expressing their intra-individual differences and forming multiple perspectives and a broad view of the concepts in focus.

As regards the ways in which students collaborated within their groups, it is worth noting that students articulated and shared their ideas with their colleagues through face-to-face communication, so as to form complete and correct solution strategies to the 1st task. As regards the role of whole class discussions, it is worth noting that, each student shared and articulated the specific solution strategy that he/she proposed in their group with the whole class. Each student also provided the whole class with detailed replies to their questions.

6 Discussion

As is clearly demonstrated by the data presented in this case study, the various tools provided by Cabri inspired each group to use them in a diversity of ways to meet the demands of the 1st task, which was for it to be solved 'in as many ways as possible'. As a result, each student expressed - within their own group - their perspectives on the concepts in focus in terms of his/her own solution strategies to the tasks at hand. With students articulating and negotiating their solution strategies within their groups and in front of the whole class, each student also developed multiple perspectives on the said concepts, which in turn helped them to construct multiple solution strategies to the 2nd task, which also required being solved 'in as many ways as possible'. Because both tasks were required to be solved 'in as many ways as possible', the students during both team and individual work were encouraged to invent various solution strategies. In this way, students developed multiple perspectives on the issues at hand at both group and personal level. The possibility of receiving multiple feedback from Cabri and from their collegues during group work and whole class discussions also gave the students confidence in the correctness of the solution strategies they constructed.

Based on the promising results to emerge from the use of MULTIPLES in the reported face-to-face communication setting, we can predict that similar encouraging results could emerge from the use of MULTIPLES within web-based settings. In fact, within such learning settings, students can be provided with fruitful collaboration support in terms of: (a) synchronous discussion capabilities, such as chat rooms - for each group and for the entire class - and videoconferencing. With the provision of such possibilities, students have the chance to discuss case issues that require real time brainstorming and sharing of information, debate controversial issues and plan their actions in a short time, (b) asynchronous discussion opportunities using bulletin boards and discussion forums. By participating in online discussions, students can articulate their understanding of the issues at hand by answering their colleagues' questions and explaining their points of view. Students also have the opportunity to hear other students' viewpoints, ask questions and revisit these discussion areas - in their own time - to reflect on their postings. On the whole, in the context of online discussions, students can collaborate and negotiate their diverse points of view on the topic at hand, while at the same time attaching multiple perspectives to it, (c) various links to multiple learning tools, search engines and learning materials, such as microworlds, simulations, and virtual reality environments, as well as online databases and information repositories presented, and by using multimedia, such as text, digital, audio and video. Students have the chance to form different perspectives on an issue by accessing these various links and learning materials, (d) publishing/sharing their work online (e.g. their solution strategies to the given problems). Group members can also work simultaneously and co-edit documents online and annotate them, if an appropriate application sharing tool (groupware) is provided. In this way, students can be engaged in both reflection and peer evaluation of each other's solution strategies. As a result, students can enhance their perspectives on the issue at hand.

7 Conclusions

In this study, a learning framework – the MULTIPLES framework – has been proposed that encourages collaboration within contexts providing multiple aids in terms of cognitive tools, representation systems and feedback, at the same time asking students to perform tasks in 'as many ways as possible'. The role of this framework has been examined through a case study using real students with promising results. In fact, students exploited the abundance of tools provided by the educational software Cabri-Geometry II as well as multiple ways of collaboration – within their groups and the whole class - and constructed a plethora of solution strategies to the tasks at hand at both group and individual level. Based on the data emerging from this experiment, it was clearly shown that, in supporting the generation of multiple perspectives of the tasks at hand, the role of engaging students in learning environments rich in learning tools, representation systems and communicating interactions is a crucial one. In addition, the kind of activities that can help learners express their inter-individual and intra-individual varieties is of major interest. To this end, multiple-solution activities to be performed in contexts providing a variety of tools are the most appropriate. Specifically, asking both groups and each individual to form solution strategies to the tasks at hand student *'in as many ways as possible'* provides them with the opportunity to exploit the richness of the aforementioned learning environments. Students can select from among various cognitive tools and representation systems those most relative to their knowledge to form their own solution strategies to the given tasks, at the same time expressing their intra- and inter-individual variety and acquiring different learning styles. Students can also use the provided communication capabilities to express their knowledge as well as to exploit the knowledge of their colleagues. The study data also clearly supported the notion that computer learning environments are ideal for integrating both cognitive tools and representation systems which can be combined by students to form multiple solution strategies to the tasks at hand. Web-based learning environments can also provide learners with rich communication capabilities so that they may exchange ideas, and articulate and negotiate opinions. On the whole, the enhancement of Web-based environments with appropriate cognitive tools, in tandem with the requirement for learners to solve tasks 'in as many ways as possible', provides students with great opportunities to develop multiple perspectives on the issue at hand.

References

1. Bednar, A.K., Cunningham, D., Duffy, T.M., Perry, J.D.: Theory into practice: How do we link? In: Anglin, G.J. (ed.) Instructional Technology: Past, present and future. Libraries Unlimited, Englewood (1991); Bennett, N.: Teaching styles and Pupil Progress. Harvard University Press, Cambridge (1977)
2. Cohen, L., Manion, L.: Research Methods in Education. Routledge, London (1989)
3. Collins, A.: Cognitive apprenticeship and instructional technology. In: Idol, L., Jones, B.F. (eds.) Educational values and cognitive instruction; Implications for reform, pp. 121–138. Erlbaum, Hillsdale (1991)
4. Dabbagh, N., Bannan-Ritland, B.: Online learning: Concepts, strategies, and application. Pearson, Merrill Prentice Hall, Upper Saddle River (2005)
5. Duffy, T.M., Cunningham, D.J.: Constructivism: Implications for the design and delivery of instruction. In: Jonassen, D.H. (ed.) Handbook of educational communications and technology, pp. 170–198. Simon & Schuster Macmillan, New York (1996)

6. Dabbagh, N.: Pedagogical models for E-Learning: A theory-based design framework. International Journal of Technology in Teaching and Learning 1(1), 25–44 (2005)
7. Daradoumis, T., Marques, J.M.: Distributed Cognition in the Context of Virtual Collaborative Learning. Journal of Interactive Learning Research 13(1/2), 135–148 (2002)
8. Dennen, V.P.: Task structuring for on-line problem-based learning: A case study. Educational Technology & Society 3(3), 329–336 (2000)
9. Dillenbourg, P., Schneider, D., Synteta, P.: Virtual Learning Environments. In: Proceedings of 3rd Panhelenic Conference: ICT in Education, Rhodes, Greece, pp. 3–18 (September 2002)
10. Duch, B.J., Groh, S.E., Allen, D.E.: Why problem-based learning? A case study of institutional change in undergraduate education. In: Duch, B.J., Groh, S.E., Allen, D.E. (eds.) The Power of Problem-Based Learning, pp. 3–12. Stylus Publishing, Sterling (2001)
11. Duffy, M.T., Lowyck, J., Jonassen, H.D.: Designing environments for constructive learning. Springer, Berlin (1993)
12. Harasim, L.: Online Education: Perspectives on a new Environment. Praeger, N.Y (1990)
13. Harasim, L., Hiltz, S.R., Teles, L., Turoff, M.: Learning Networks: a field guide to Teaching and Learning Online. MIT Press, Cambridge (1995)
14. Hart, K.-M.: Measurement. In: Murray, J. (ed.) Childrens Understanding of Mathematics, pp. 9–22. Athenaeum Press Ltd., G. Britain (1989)
15. Janvier, C.: Representation and understanding: The notion of function as an example. In: Janvier, C. (ed.) Problems of representation in teaching and learning of mathematics, pp. 67–72. Lawrence erlbaum associates, London (1987)
16. Jonassen, D.H.: Objectivism versus constructivism: do we need a new philosophical paradigm? Journal of Educational Research 39(3), 5–14 (1991)
17. Jonassen, D.H., Carr, C., Yueh, H.-P.: Computers as Mindtools for Engaging Learners in Critical Thinking. Tech. Trends 43(2), 24–32 (1998)
18. Jonassen, D.H.: Designing constructivist learning environments. Instructional design theories and models 2, 215–239 (1999)
19. Jonassen, H.D.: Revisiting Activity Theory as a Framework for Designing Student-Centered Learning Environments. In: Jonassen, D.H., Land, S.M. (eds.) Theoretical Foundations of Learning Environments, pp. 89–121. Lawrence Erlbaum Associates, Mahwah (2000)
20. Joyce, B., Weil, M.: Models of Teaching. Prentice Hall, Englewood Cliffs (1972)
21. Kaput, J.J.: The Representational Roles of Technology in Connecting Mathematics with Authentic Experience. In: Biehler, R., Scholz, R.W., Strasser, R., Winkelman, B. (eds.) Didactics of Mathematics as a Scientific Discipline: The state of the art, pp. 379–397. Kluwer Academic Publishers, Dordrecht (1994)
22. Kordaki, M.: The effect of tools of a computer microworld on students' strategies regarding the concept of conservation of area. Educational Studies in Mathematics 52, 177–209 (2003)
23. Kordaki, M.: The role of multiple representation systems in the enhancement of the learner model. In: 3rd International Conference on Multimedia and Information and Communication Technologies in Education, Caceres, Spain, June 2005, pp. 253–258 (2005)

24. Kordaki, M.: Multiple Representation Systems and Students' Inter-Individual Learning Differences. In: Kommers, P., Richards, G. (eds.) Proceedings of World Conference on Educational Multimedia, Hypermedia and Telecommunications (ED-MEDIA), pp. 2127–2134. AACE, Chesapeake (2006)
25. Kordaki, M., Balomenou, A.: Challenging students to view the concept of area in triangles in a broader context: exploiting the tools of Cabri II. International Journal of Computers for Mathematical Learning 11(1), 99–135 (2006)
26. Laborde, J.-M.: Cabri-Geometry. Universite de Grenoble, France (1990)
27. Lemerise, T.: On Intra Interindividual Differences in Children's Learning Styles. In: Hoyles, C., Noss, R. (eds.) Learning Mathematics and Logo, pp. 191–222. MIT Press, Cambridge (1992)
28. Maureen, T.: Constructivism, Instructional Design, and Technology: Implications for Transforming Distance Learning. Educational Technology & Society 3(2), 50–60 (2000)
29. Murphy, R., Ness, G., Pelletier, J.: Problem-Based Learning Using Web-Based Group Discussions: A Positive Learning Experience for Undergraduate Students. In: Crawford, C., et al. (eds.) Proceedings of Society for Information Technology and Teacher Education International Conference 2001, pp. 1149–1154. AACE, Chesapeake (2001)
30. Nardi, B.A.: Studying context: A comparison of activity theory, situated action models, and distributed cognition. In: Nardi, B.A. (ed.) Context and consciousness:Activity theory and human-computer interaction, MIT Press, Cambridge (1996)
31. Noss, R., Hoyles, C.: Windows on mathematical meanings: Learning Cultures and Computers. Kluwer Academic Publishers, Dordrecht (1996)
32. Papert, S.: Mindstorms: Pupils, Computers, and Powerful Ideas. Basic Books, New York (1980)
33. Pea, R.D.: Practices of distributed intelligence and designs for education. In: Salomon, G. (ed.) Distributed cognitions. Psychological and educational considerations, pp. 47–87. Cambridge University Press, NY (1993)
34. Rogoff, B.: Developing understanding of the idea of communities of learners. Mind, Culture, and Activity 4, 209–229 (1994)
35. Rogers, Y.: A Brief Introduction to Distributed Cognition (August 1997), http://www.cogs.susx.ac.uk/users/yvonner/dcog.html (Retrieved on September 11, 2008)
36. Salomon, G.: Distributed Cognitions: Psychological and educational considerations. Cambridge University Press, New York (1993)
37. Schmeck, R.R.: Learning Strategies and Learning Styles. Plenum, New York (1988)
38. Silver, E.A., Leung, S.S., Cai, J.: Generating multiple solutions for a problem: a comparison of the responses of U.S. and Japanese students. Educational Studies in Mathematics 28, 35–54 (1995)
39. Snow, R.E.: Individual differences and the design of Educational programs. American Psychologist 41, 1029–1039 (1986)
40. Vygotsky, L.: Mind in Society. Harvard University Press, Cambridge (1978)
41. Wilson, B.G., Cole, P.: Cognitive teaching models. In: Jonassen, D.H. (ed.) Handbook of research for educational communications and technology, pp. 601–621. Simon & Schuster Macmillan, New York (1996)

E-Learning at Work in the Knowledge Virtual Enterprise

Nicola Capuano[1], Sergio Miranda[2], and Francesco Orciuoli[1]

[1] DIIMA – Department of Information Engineering and Applied Mathematics University of Salerno, Via Ponte don Melillo 84084 Fisciano (SA), Italy
[2] MOMA – Mathematical Models and Applications, via Marcello 2/6, 84085 Mercato S. Severino (SA), Italy

Abstract. The purpose of this chapter is to propose an overview of the Knowledge Virtual Enterprise model, where the Virtual Enterprise vision is extended with Knowledge-based assets in order to provide an agreement model to support the interoperability among organizations. Every enterprise or organization, by itself, is a source of original knowledge that, if exploited, can contribute to its competitiveness. If this is true inside the enterprise walls, it is more relevant when extended to Virtual Enterprises, especially when they operate in a tumultuous and unsettled context, like ICT, strongly bound to the so called soft skills and even more to the capability of carrying out just-in-time knowledge take-over and transfer. In order to explain the advantages of the Knowledge Virtual Enterprise model we define some real-world business scenarios, to be executed within the context of a Knowledge Virtual Enterprise instance. The scenarios are based on the idea that several organizations could put together their competences, human resources, expertise, technologies, etc. to carry out complex project activities, requiring resources that are usually difficult to be found in a single organization. The scenarios are particularly focused on how the Knowledge Virtual Enterprise model can support personalized, contextualized, effective and efficient e-learning at work experiences. Finally, the Knowledge Virtual Enterprise model vision is concretized through the description of a feasible technological mapping between its main concerns and existing software technologies and specifications.

1 Introduction

e-Learning is usually identified as a formal piece of learning or specialist software, but the simple use of a computer frequently means involving someone in formal or informal e-learning processes: looking for subjects in internet, watching online tutorials and interacting with something or, simply, communicating.

e-Learning at work was born few years ago when Ministries and Institutes (for example the National Institute for Continuing Adult Education in UK) looked at the way to involve adults in e-learning programs. By doing this, they found the key challenge: how to link employees' personal skill needs with organizational learning and development processes.

To describe what really e-learning is in the workplace we should mention three trends. The first one is the internet-based educational model (by using of "learning management systems" and "learning content management systems"). The second one is the classroom model in learning designs (often identified as "informal learning" or "work based learning"). The third one is based on exchange of experience and development of knowledge. Many of these models evolved to the "socio-cultural approach" to human learning, by means of "practice communities" and "collaborative construction" of knowledge.

e-Learning at work could mean using these evolved models and coordinating them with all HR management as well as management processes so to develop an organizational climate and social and physical environment beneficial to organizational learning.

Thus, e-learning programs should be quite easily used to teach a wide range of skills, job-specific skills able to give professional qualifications.

Although employers are keen to learn more about e-learning and its business impact, they would like to make better use of interactive software, receive well personalized e-learning path and be involved in programs very closed to their activities, useful to improve their ranking or receive rewards.

Aware of the objectives of e-learning at work [1], [2], this chapter tries to describe a possible approach based on a Knowledge Virtual Enterprise.

2 Knowledge in the Virtual Enterprises

A Virtual Enterprise (VE) may be considered as a temporary association of autonomous and heterogeneous enterprises (for instance a network of SMEs), whose aim is the building of ties and relationships that can be easily translated into business processes to gain profit [3].

Every enterprise or organization, by itself, is a source of original knowledge that, if exploited, can contribute to its competitiveness. It frequently happens that a solution developed in a department could be reused with benefit by another department or these ideas that are not useful in a particular moment become successful in the future. If this is true (and someway managed, with some limitations, by the current enterprise Knowledge Portals) inside the enterprise walls, it is more relevant when extended to VEs, especially when they operate in a tumultuous and unsettled context, like ICT. This sector is indeed strongly bound to the so called soft skills (relational and communication capabilities, knowledge transmission, management, organization) and even more to the capability of carrying out just-in-time knowledge take-over and transfer.

A Virtual Enterprise is able to efficiently react to change only through reusing and synergetic combination of knowledge that can be found in different nodes of its structure.

The conceptual model of the Knowledge Virtual Enterprise is built around three main roles that can give life to different applicative scenarios:

- **KBroker:** who receives a request for information retrieval, looks for the provider that is likely to be able to satisfy the request, contacts the provider and forwards the request.

- **KProvider:** who supplies resources and services to be shared in the network.
- **KConsumer:** the user of resources supplied by providers.

A layered conceptual view of the Knowledge Virtual Enterprise (KVE) model is depicted in Figure 1.

The layers of the KVE model Conceptual View (as illustrated in Figure 1) are:

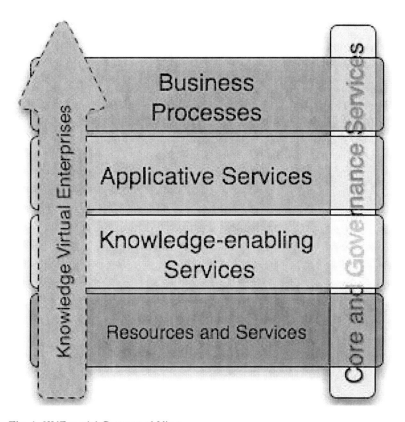

Fig. 1. KVE model Conceptual View

- **Core and Governance Services** collect services useful to a manager to administer all dynamic aspects that can occur during a KVE lifetime, such as Membership Services and Policy and Enforcement Services. Core and Governance Services are deployed towards the IT infrastructure of every organization subscribing KVE rules.
- **Resources and Services** of single organizations are exposed through a virtualization process realized following the rules defined by the KVE. An organization could expose, for instance, its human resources (skill, availability, salary, ...), its products (price, availability, features, ...), its services (e.g. on-line assistance through an automatic responder).

- **Knowledge-enabling Services** must:
- realize a semantic correlation among resources and services exposed by organizations belonging to the KVE (e.g. offering functionalities to implement KBroker, KProvider e KConsumer roles)
- manage knowledge that has been generated and used during a KVE lifecycle
- model, through taxonomies and ontologies, the KVE structure and processes.

- **Applicative Services** belong to the set of services, eventually based on Knowledge-enabling Services, operating on both the "horizontal" level (e.g. Business Process Definition and Execution, Project Planning, Document Sharing services) and the "vertical" level (e.g. e-learning, e-commerce services).

The following picture (Figure 2) shows a schema of a possible configuration (network deployment) of a KVE. The KVE, from the network deployment point of view, is represented by a graph, whose nodes are the different organizations.

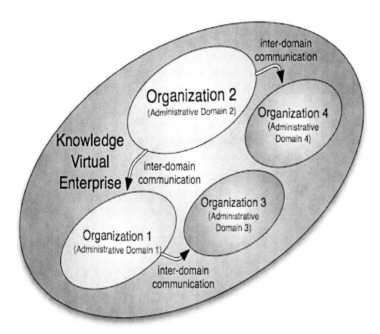

Fig. 2. A sample KVE configuration for four organizations

The KVE model could not be applied in the real world if the involved organizations do not share:

- the same representation model for their internal structures (strategies, mission, relevant case history, in progress activities, etc.).
- the same representation model for their applicative domains.
- the same representation model for their processes.

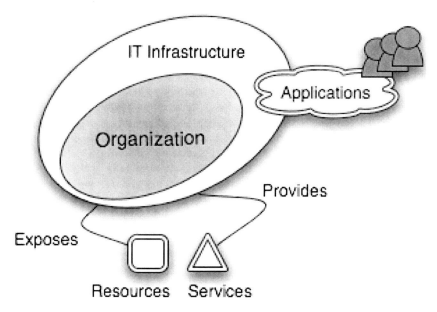

Fig. 3. Every organization exposes resources and services through an IT infrastructure

Semantic structures like taxonomies and ontologies are used to formalize knowledge on a specific domain and on processes definition. In particular, for ICT domain (adopted as the reference domain in this work), Knowledge-enabling Services manage an Enterprise Ontology (model used by ICT enterprises to expose their internal structure) [4], an ICT Process Ontology (used to identify components needed for processes definition - like a software development process) [5], and an ICT Taxonomy (model used to structure skills related to ICT domain and allow therefore expertise identification across different organizations).

3 E-Learning at Work: Some Business Scenarios

The Ministry of Agriculture, with the goal of keeping up development and innovation, announces a notice of competition for the realization of a platform to support and manage the supply chain of agro industrial sector.

Bob's company, specialized in Engineering in manufacturing industries, after downloading the tender document from the Ministry web site, considers this job interesting as an incentive for its core business (process management) development in a different context with respect to previous experiences. To find new business opportunities it is sometimes required to be able to explore new directions without loosing competitiveness.

Since the company size and expertise are not sufficient for the complete satisfaction of the competition requirements, it is needed to find partners, interested in the specific business, to cooperate with. The company management carries on a selection among the known companies, to find the ones having a good chance to be interested in the competition and starts a negotiation to verify the feasibility of the cooperation. This negotiation activity has a successful ending with two of the involved companies: Alice's company and Fred's company. An agreement is then entered among the three companies, to establish a temporary pool of enterprises to work out a project proposal in order to answer to the Ministry competition notice. Bob's company, acting as group leader, delivers, before the deadline, the complete required documentation.

The Ministry committee evaluates the presented proposals and awards the project realization to the consortium established by Bob, Alice and Fred companies. Bob's company, which has operated for the consortium establishment, is in charge of the project management, and for this reason chooses one of its employees who is eligible to act as a project manager, and formally appoints him. The Project Manager carefully analyses the competition tender, and works out a Work Breakdown Structure (WBS), to improve the project management and control through the hierarchical division of the project itself into its components and activities.

Next step is the assignment of responsibilities on all the activities, through the definition of an Organization Breakdown Structure (OBS), which is a hierarchical division of the project responsibilities, with the goal of a unique definition of resources assigned to different tasks, and the building of a Responsibility Assignment Matrix (RAM), expressly involving the consortium organizations according to the rules defined in the agreement.

At this stage a recurring problem in ICT domain projects management is perceived: no resources with enough expertise are available in the organizations belonging to the pool. As it often happens, due to the needs of the companies to answer to the always changing situations in a dynamic context, there is a gap between the required and available skilled resources. The decision is therefore to create a training project customized and tailored on the problem to be faced and to involve additional resources.

For every resource a training plan is prepared, considering the comparison between the competences owned and the skills required in order to participate in a profitable way to the project. But this is not enough to have a final training plan: very often the choices of the involved companies can take different directions and influence the process, depending on the criticality and on the value perceived concerning the competences to be acquired and the results to be obtained; Fred's company is a "young" company, its employees have no significant working experience, so intensive training during office hours is preferred, to shorten learning process. Bob's company on the contrary can rely on skilled staff, and has the need to specialize a few people on a particular topic, but with different levels of expertise to reach and different availability in terms of hours, due to their involvement in other in progress activities.

Furthermore, resources allocated to the project have different kind of studies experiences, different certifications, and different learning speed. One of them has serious visual handicap. For these reasons, customized learning projects are designed for each worker. The involved resources participate to the training project, following the planned methods and timings and go through a final assessment test aimed to estimate the progress done, to certify the adequacy of skills acquired with respect to the foreseen activities.

At this stage the Project Manager of Bob's company organizes a first workshop aiming to share the results obtained up to now and to officially start the production activities.

At the same time, a high school is starting training and learning activities to involve students in scientific experiments, but, indeed, it haven't adequate laboratory equipment to employ for motivating experiments.

The Principal of the school has to realize this project and so it asks for collaboration from technical agencies specialized in virtual scientific experiments execution. After a negotiation activity, two agencies are selected to work out the project. One of them has a powerful simulator able to create virtual environment where it is possible to study the planets motions, to reproduce the interactions among masses in gravity field, to observe trajectories and reactions. Their researchers know very well all details of these phenomena and related simulations. The other agency has an engine able to design complex mathematical models related to fluid dynamic phenomena. Also the researchers working in this agency know very well all details about that.

In both cases practical and institutionalized knowledge are available. The collaboration with these organizations allows the school to improve the efficiency and efficacy of its learning and training activities. In the first step the teachers of the schools may receive support from the knowledge of the experts of the agencies. In the second step, young students may be engaged in simulations and get in virtual laboratories where they may apply all competencies acquired during the normal lessons, where they may find all knowledge shared by the teacher and the experts, where they may receive detailed and personalized deepening as they could need.

4 How a Knowledge Virtual Enterprise Leverages E-Learning at Work

In this section we are going to see how the scenarios introduced in the previous section could be realized using the KVE model.

To begin, we suppose that a KVE Administration Application Node (KAAN) and a KVE Knowledge-enabling Services Node (KKSN) are available and can be accessed through the network. The KAAN allows to access the functionalities of lifecycle management of a KVE, while the KKSN, beyond other functionalities, exposes a registry service to record the references to all the organizations adopting the model of the KVEs.

Through the KAAN and the KKSN, Bob (who represents his organization) looks for partners to be involved in the activities foreseen by the requirements expressed by the Ministry of Agriculture. The KKSN answers to Bob's query, and returns as a result (using the KAAN front-end) information on Alice's organization and Fred's organization. Once retrieved the references of the two organizations, Bob starts the KVE and a process (in charge of the Business Process Execution Service exposed by the KBPN) with the goal of producing an agreement. After this first task, the KVE may have a new goal: in a collaborating way the involved organizations have to produce and submit the project proposal. When to proposal is accepted, Bob decides his resources to employ for the project and, by using KBPN, define the project workflow. For each activity, he also asks to the other companies to allocate human resources. The request is carried out considering skills and characteristics expressed through a Competency Model and a Worker Model [6]. All the requests referring to the resources needed for project activities are registered by the KKSN.

Once the human resources have been chosen, the processes related to project activities can be activated and therefore run by the KBPN. The KVE is also used to fulfill the skill gap, through the execution of learning experiences personalized and customized on the needs of learners (workers in this case) and adapted to the project context. These services could be deployed on a new network node that is the e-Learning Services Node (ELSN). The ELSN, therefore, receives information about competences of the involved workers/learners and the project requirements and builds a learning path on the best set of learning objects from several Digital Repositories useful for the acquisition of the needed competences. At the end of the learning experience, workers will go through assessment tests and then to the activities if they pass the test or to a remedial learning path.

In the same time, in the other scenario, the Principal of the school looks for expert agencies, through the KAAN and the KKSN, he receives information from two agencies able to create and execute virtual scientific experiments. He creates a KVE with the selected agencies and define the goal to design virtual scientific experiments for his school. He find the right experiments by consulting with the expert of the agencies in a collaborative way and start a new business processes for each of them by the Business Process Execution Service exposed by the KBPN. He allocates right teachers and select young students considering their skills, competencies and real needs expressed through a Competency Model and a Learner Model. All the requests referring to the resources needed for activities are registered by the KKSN. By using the ELSN, all involved users are get in customized learning and training activities: teachers may receive personalized support from the agencies and access to their shared knowledge; students may receive personalized learning and training path to be able to assist to a virtual scientific experiment, to interact with it, to understand which phenomenon it represents.

5 Technological Mapping

In this section, our effort is the identification of the existing technologies and specifications that could be exploited to implement in real terms the previously described scenarios, and establish a set of foundations to support design and implementation of a software platform for the KVEs life cycle management. The KVE model defined in this chapter is tightly-coupled with the concept of "service", so we believe that we cannot avoid to use concepts coming from the SOA model [7], actually considered at the state of the art for the distributed architectures. Basing on SOA concepts we identify two main critical points (there exist many other critical points) for the implementation of our software platform:

- **Execution of business processes** across several administrative domains. This problem is strictly correlated to the difficulties of localizing, instantiating and orchestrating services deployed in multiple administrative domains. The difficulties are due to the different policies, technologies, etc. used by every administrative domain. The solution for the previously described problem seems to be the Grid Technology. Globus Toolkit 4 [8] and WSRF (Web Service Resource Framework) [9] are middlewares (implementing the SOA model) that provide a framework for the development of the so called Grid Services (i.e. services implemented on the top of Web Service technology and whose life cycle can be managed within a Grid Middleware) and a set of built-in services able to govern the life of Grid Services. The Grid Middleware enables the setting of Virtual Hosting Environments in which service clients can use transparently services deployed in their own administrative domain, as well as services deployed in a foreign administrative domain. Security, Privacy and Policy Management are other important issues related to the execution of business processes [10]. WS-Security, WS-Privacy, WS-Trust and WS-Policy represent a subset of the existing specifications addressing the aforementioned issues [11][12]. Furthermore, with respect to the service orchestration and the business process execution, there exist some formal languages like BPEL (its extension WS-BPEL [13] addresses the orchestration of Web Services) and BPEL4PEOPLE (that enables the definition of human interactions with the activities of a workflow).

- **Re-use of existing IT infrastructure.** The real-world rules say to us that we cannot re-implement from scratch all the services that organizations would like to expose in the context of a KVE. So, we need to re-use more and more existing IT infrastructures available in enterprises. Once more the Grid Middleware provides to us virtualization mechanisms able to hide complexity and peculiarities of existing legacy applications, databases, services, etc. and expose only their functionalities using standard protocols and specifications. Grid Services represent a good "wrapping mechanism" solution for legacy systems. Using Grid Services we can virtualize organization-specific resources and services and plug them into KVE environments. So, consumers, using standard mechanisms, can exploit virtualized resources and services. In the defined scenario, organizations have to expose human resources competences. Nowadays, enterprises choose a Human Resource Management System (HRMS) with respect to their strategies, their budgets and their requirements. SAP ERP HCM, Oracle's PeopleSoft Enterprise and Microsoft Dynamics GP/AX are only few samples of the existing HRMS solutions. SAP, Oracle and Microsoft HRMS have, in general, different ways to manage human resources and different ways to represent human resources competences. In order to plug all of the aforementioned systems into a KVE instance we have to virtualize the way they represent and provide access to human resources competences. In this case, the virtualization process can be realized in two main stages: (a) adhering to a shared vocabulary in order to identify names and semantics of all relevant competences in the ICT world; (b) using the same competences representation format in order to support interoperability across organizations within a KVE. The shared vocabulary in the KVE model is provided by the ICT Taxonomy, that is managed by Knowledge-enabling Services. The common competences representation format could be provided by HR-XML [14]. HR-XML is a XML specification designed to enable e-business and the automation of human resources-related data exchanges. The HR-XML specifications developed by the HR-XML Consortium define the interface necessary to automate background investigation orders and enable fast, easy access to background investigation reports. In the end, we would like to spend some words about the two domain-specific Applicative Services used in the defined scenario: Skill Gap Analysis and Automatic Unit of Learning Building services. In order to provide the implementation of the aforementioned services we could use the IWT (Intelligent Web Teacher) e-learning platform [15], able to offer personalized and contextualized learning experiences exploiting e-learning ontologies in order to represent educational domains knowledge. The pieces of interest of the ICT Taxonomy can be simply mapped onto a IWT-compliant ontology. IWT also provides a mechanism to represent cognitive states and learning preferences of learners using an e-learning standard called IMS-LIP [16]. Competences represented in HR-XML could be mapped (in a simple way) onto a IMS-LIP structure.

6 Conclusions and Future Works

In this chapter, we have presented the Knowledge Virtual Enterprise model, where the management of the virtual enterprises lifecycle is supported by a set of Knowledge-based resources and services. To foster the proposed model, we have defined some real-world scenarios in which several organizations, with different policies and different IT infrastructures constitute a Knowledge Virtual Enterprise with the aim of proposing, planning and executing project activities. The proposed scenarios are focused on the issue of e-learning at work and in particular on the assembling and provisioning of personalized and contextualized learning experiences for effective and efficient training on the job activities. The personalized and contextualized learning paths outcome from the skill gap analysis between human resources (allocated onto project activities) skills and required skills (needed to carry out project activities). Human resources skills are exposed by organizations in the context of the KVE, while, required skills are provided by the ICT Taxonomy. Furthermore, we have identified a set of technologies and specifications useful to implement the Knowledge Virtual Enterprise model and in particular the proposed scenario. The technological mapping has revealed (a) Grid Middleware as foundations to build Knowledge Virtual Enterprises, and (b) a set of specifications as mechanisms to solve specific problems, like business processes definition and execution, security/privacy/policy, interoperability, etc.

Future works on the vision proposed in this chapter consist of a definition of a detailed software architecture that implements the Knowledge Virtual Enterprise model.

References

1. Bélanger, M.: Work-based distributed learning. In: Encyclopedia of Distributed Learning. Sage Publications, Thousand Oaks (2004)
2. Bersin, J.: Making rapid e-learning work. Chief Learning Officer (July 2006), http://www.clomedia.com/content/templates/clo_article.asp?articleid=1008&zoneid=6 (Retrieved April 2006)

3. Camarinha-Matos, L.M., Afsarmanesh, H.: Elements of a base VE infrastructure. J. Computers in Industry 51(2) (June 2003)
4. Uschold, M., King, M., Moralee, S., Zorgios, Y.: The Enterprise Ontology The Knowledge Engineering Review. In: Uschold, M., Tate, A. (eds.). Special Issue on Putting Ontologies to Use, vol. 13 (1998)
5. Jenz, D.E.: Business Process Ontologies: Speeding up Business Process Implementation, BPTrends (2003) (November 2004)
6. Sicilia, M.A.: Ontology-based competency management: Infrastructures for the knowledge- intensive learning organization. In: Lytras, M.D., Naeve, A. (eds.) Intelligent learning infrastructures in knowledge intensive organizations: A semantic web perspective, pp. 302–324. Idea Group, Hershey (2005)
7. Foster, I., Kesselman, C., et al.: The Anatomy of the Grid Enabling Scalable Virtual Organizations (2001)
8. Dimitrakos, T., Mac Randal, D., Yuan, F., Gaeta, M., Laria, G., Ritrovato, P., Serhan, B., Wesner, S., Wulf, K.: An Emerging Architecture Enabling Grid Based Application Service Provision. In: Proceedings of the 7th International Conference on Enterprise Distributed Object Computing, September 16-19, 2003, p. 240 (2003)
9. Humphrey, M., Wasson, G.: Architectural Foundations of WSRF.NET (2005), http://www.cs.virginia.edu/~gsw2c/wsrf.net.html
10. Wasson, G., Humphrey, M.: Toward Explicit Policy Management for Virtual Organizations. In: Proceedings of the 4th IEEE International Workshop on Policies for Distributed Systems and Networks, June 04-06, 2003, p. 173 (2003)
11. Web Services Security OASIS Standard Specification (2006), http://www.oasis-open.org/committees/download.php/16790/wss-v1.1-spec-os-SOAPMessageSecurity.pdf
12. Web Services Policy Framework, http://www-128.ibm.com/developerworks/library/specification/ws-polfram/
13. Web Services Business Process Execution Language Version 2.0, http://www.oasis-open.org/committees/tc_home.php?wg_abbrev=wsbpel
14. HR-XML Consortium, http://www.hr-xml.org
15. IWT Intelligent Web Teacher, http://www.didatticaadistanza.com/
16. IMS Learner Information Package Specification, http://www.imsglobal.org/profiles/index.html

Communities of Practice Environment (CoPE): Democratic CSCW for Group Production and E-Learning

David Thaw[1,2], Jerome Feldman[1,2], Joseph Li[2], and Santi Caballe[2]

[1] University of California, Berkeley
  School of Information
  102 South Hall
  Berkeley, CA 94720
  USA
[2] International Computer Science Institute
  1947 Center St.
  Suite 600
  Berkeley, CA 94704
  USA
  {dbthaw, feldman, joeli, scaballe}@icsi.berkeley.edu

Abstract. Computer Supported Cooperative Work (CSCW), Democratic Group Production and e-Learning have overlapping but not identical design and use profiles. We suggest that democratically organized CSCW systems provide an ideal platform for organizing and encouraging group interactions in e-Learning and other democratic group production situations where physical meetings are not feasible. Few attempts at democratic CSCW systems have been made, we argue, largely because the necessary development resources are concentrated in large hierarchical organizations which have little apparent use for such systems. The CoPE CSCW system does seem to provide a promising platform for extension to e-Learning and other characteristically similar environments, and there is an ongoing effort to do this.

1 Introduction and Background

Democratic decision making is a principle in which the content and management of workflow operates through some form of group-based discussion and voting mechanisms. There are a variety of potential voting mechanisms, as evidenced in mainstream political science literature [1]. The CoPE system introduced in this paper distinguishes itself by employing some of these methods to control how a document transitions through the system workflow. In a classic workflow example, draft documents may be commented upon in a given workflow state (deliberation and democratic participation); after the commenting period the document may move into a decision state during which a vote takes place to determine if the document will be approved or rejected (democratic decision). Democratic production of documents is related to the first characteristic, but focuses on the process by which documents are created rather than that by which they are accepted by a group.

Collaborative learning is an emerging paradigm dedicated to improving teaching and learning with the help of modern information and communication technology [2]. Democratic group-based support can facilitate both formal and informal learning, particularly increasing the efficacy of informal learning structures. Modern pedagogical approaches include advanced learning techniques based on some form of collaborative consensus-building mechanism, such as learning by discussion and problem-based learning [3]. Additionally, such mechanisms can substitute for the lack of a central knowledge authority common to informal groups such as non-technical students collaborating in small groups outside formal classes.

This paper focuses on the intersection between collaborative work (CSCW) [4], collaborative learning (CSCL), [5] and Document-Centric Democratic Collaboration and Production (DCDCP) to help small groups – with or without technical skills – to enhance and improve the construction of effective knowledge [6].

Questions related to the application of DCDCP to small knowledge groups have been, to the best of our knowledge, the subject of limited investigation in online spaces. Software development efforts have been primarily focused on commercial applications for business customers. These efforts focused on enterprise-scale applications, although some projects were targeted to smaller groups such as small businesses or business units within larger organizations. Even those targeted at smaller environments, however, generally envisioned commercial users and none considered the type of democratic collaborative production and enabling of group *formation* that is needed in our research context.

At the beginning of our research, we conducted an initial application survey [7] as an attempt to examine ways in which communities with common interests have collaborated online in exchanging information or building knowledge repositories of some form. We examined a wide cross-section of interest groups that share information and benefit from collaborative learning. The composition of such groups ranged from online associations of networked professionals, short-term collaborators on specific projects, businesses, hobby groups, and social exchange networks. Several factors with wide variance included: group size, duration of participation, proximity of individual participants, economic relationship between participants, and level of intra-group interactivity.

Our initial survey found little activity in group-based e-Learning of a DCDCM style. Thus we broadened our search and conducted a more formal review of the current state-of-the-art online collaborative systems most likely to operate in this space, either currently or that might develop in this direction in the future. While we found several approaches in this direction [4], [8], four essential characteristics in the DCDCM domain were missing: 1) democratic decision making; 2) democratic production of documents; 3) usability of the system by general users; and 4) customizability of the system by general users.

In this work, an attempt to fill this gap is reported by introducing an innovative system called Communities of Practice Environment (CoPE).

2 Introduction to CoPE

The fundamental idea behind CoPE [9] is the result of both an inductive and iterative deductive investigation process. The inductive process began with democratic theory and CSCW first-principles as described in Section 1. From this initial loose specification we worked deductively based on our experience with several democratically-organized groups. Examples of these groups included an immigrant-support network, an interdisciplinary faculty at a university, an academic conference, and a technology-based education program for secondary school students. We envision this type of system enabling collaboration for other types of groups such as social workers, community action groups, public school teachers, and other sets of individuals who might benefit from the ability to organize and collaborate outside their traditional institutional structures.

CoPE is designed to enable a specific type of collaboration; a subset of CSCW that we argue has not been adequately addressed. Specifically, this involves sets of individuals who share a need or desire to engage in collaborative production. The object of this production is something that can be codified in documents. CoPE is targeted to individuals who do not already have a formal workflow for this collaboration or who are seeking to improve upon inefficient workflows. CoPE also envisions enabling collaboration among individuals who are part of organizations with formal collaboration mechanisms, but whose mechanisms are limited to intra-organization collaboration. Finally, CoPE is designed to enable *collaboration*, not *management*, and thus envisions "democratic" collaboration.

There are many examples of sets of individuals around the world who have a need or desire to collaborate but lack the resources, knowledge, or institutions to do so [10]. Consider, for example, public school teachers, social workers, and community action groups (where the group and its peer groups are the "individual"). Often these individuals are separated by geography and/or time. They could be too distant from one another to organize face-to-face meetings. They also could be unable to meet due to scheduling constraints or differing work hours. Such individuals may already be part of existing organizations but the "peers" with whom they wish to collaborate are in different organizations. CoPE is especially targeted to the individuals and organizations described here who lack substantial technical expertise or the resources to acquire such expertise. This includes any e-Learning situation for non-technical students.

3 CoPE Development

In this section, we present the main guidelines that influenced the construction of CoPE. To this end, we first describe the user requirements that motivated its development and then we provide the main decisions made in designing the user interface. Finally, we discuss certain technology issues in building and supporting CoPE.

3.1 General Requirements and Analysis

CoPE was developed for the needs of a certain type of user forming the CoPE User Community. The system interface design makes assumptions based on the characteristics of such users. We call this type of user the "General User." The following assumptions motivate this definition:

- users do not have specialized (information/computing) technical skills,
- users possess a basic skill set for computer and internet usage,
- users posses the ability to learn a new (information/computing) skill set of this same basic technical level,
- users are *willing* to learn a new (information/computing) skill set of this same basic technical level,
- users do not already share a sophisticated and/or long-used method for electronic collaboration.

The general user is not a complete definition of a user's characteristics. Rather, it is a set of characteristics – both affirmative and negative – that are necessary and sufficient conditions for a general user. A general user may have many or no other characteristics, and other than meeting all the conditions there is no assumption that two general users are alike.

We conceptually framed our target user from the characteristics of the sets of individuals for which we wish to enable group formation and collaboration. As described in Section 2, we employed both inductive and deductive methodology. We began by examining the types of collaboration we wanted to enable and the types of groups we wished to facilitate forming. This generated our first condition, that users did not have specialized (information/computer) technical skills. We then considered the constraint of assuming that individuals would not necessarily be using the same computer operating platform, thus requiring a platform independent system. This led to the selection of a web-based application and thus the second condition, an affirmative requirement that users have at least a certain basic set skills in computer and internet usage.

As discussed in Section 3.2, we opted to design the user interface ourselves rather than extending a high-distribution interface[1] such as Yahoo's or Microsoft's web portal. This decision introduced the affirmative constraint that users have the ability and willingness to learn a new skill set. We assumed that our target users would not have prior experience with the Zope/Plone interface and therefore our extension of it would necessarily introduce a learning curve. We found that the ability to learn a new computing skill set, while a workable restriction, did actually exclude some users. Our anecdotal evidence suggests that age is the most common factor in determining membership in this category. Willingness to learn a new skill set, we found, was equally important[2] and our anecdotal evidence also suggests a correlation with age.

---

[1] In this context, an interface/framework with such wide distribution (see subsequent examples) that it would be a "household name" in most of our target population's vocabulary.

[2] And probably equally determinative, but it was difficult to test for the difference between inability and unwillingness with such a small sample not even designed to be a sampling population.

Deductive reasoning based on our experiments with system design test groups also produced an additional constraint. For CoPE to be successfully adopted, users often needed to "lack" a certain degree of technical sophistication. More specifically, it introduced a negative constraint against a user having existing sophisticated methods for collaboration. This was because users who had preexisting sophisticated methods, particularly ones in use for some time, found the transition costs to a new system[3] to be too high and would frequently revert to their usual methods of collaboration.

In conjunction with these early requirements and in order to both gather the specifics needs of our targeted audience and identify the lack of technological support to meet these needs, we deployed a series of test groups using then-current online collaboration tools. The collaboration tools were primarily designed for small-medium sized businesses or business units/groups within large organizations.

Our main test group was the California Central Valley Partnership (CVP). At the time the CVP group was comprised of an executive board of approximately 15 members who were leaders of public interest organizations for immigrants living and working throughout California's central valley agricultural communities. The leaders of these organizations were substantially geographically dispersed, possessed varying levels of technical skill with information systems, and communicated in multiple languages. They had few face-to-face meetings.

Our experience with the CVP confirmed much of what we expected about the products available at the time. While likely effective for their target audience, the unique characteristics of the General User and the CoPE Community made it difficult for members of the CVP to effectively use the tool. Most notable in this early experiment was the difficulty users with less technical skill had in learning a system whose interface pre-supposed a higher level of technical skill than that which we ascribe to the General User. This steep learning curve provided a strong disincentive to members of the CVP, and thus our first test resulted in negligible adoption and no performance advantage for the organization's activities.

Building on this experience and our own research, we began a requirements analysis. The CVP has remained a part of this process. As it possessed many of the characteristics of our target user base (i.e., the user CoPE community), we interviewed members of the organization, participated in some of their activities, and engaged in other similar ethnographic research methods. The results of this process were instrumental in development of our system requirements.

3.1.1 The CoPE Community

The Communities of Practice Environment, as the name suggests, is not designed to be a "one-size fits all" CSCW solution. Rather, our user research and the resulting system design are targeted for a specific type of group. These "CoPE Communities" are sets of individuals who have a common interest in collaborative

---

[3] More specifically, a new system that did not necessarily provide greater technical capabilities for collaboration. As discussed earlier, CoPE is not intended to be an improvement on collaboration systems generally; rather, it addresses a specific need that previously remained unfulfilled.

production and who lack the technical expertise or means to acquire such expertise that would enable them to implement a CSCW solution themselves.

As discussed in the section "Similar Systems and the State of the Art," CoPE is hardly unique in striving to develop an information system enabling general users to engage in productive collaboration. Our work incorporates the theories of "open" development [10] and a form of "democratic" collaboration to facilitate these individuals' finding ways to collaborate where they otherwise would be unable, regardless of their available resources. CoPE is not simply another attempt at CSCW for non-technical users. It is a means by which individuals that share nothing more than a common productive interest can organize themselves and their productivity process – while that productive collaboration is facilitated by the same CSCW system which enabled them to form their "community" in the first place.

We use the term "sets of individuals" in defining the CoPE Community because of CoPE's goal of *creating* groups. While many CoPE Communities may comprise pre-existing groups, one of the goals of CoPE is to allow collaboration – and therefore the *formation* of groups – among individuals who could not previously collaborate. The individuals who make up a CoPE Community share (at least) the following characteristics:

- they have some interest, skill, or responsibility that they share with others who could form a CoPE
- the interest, skill, or responsibility could be enhanced by and/or could enhance some productive output if they were able to collaborate with others
- they do not presently have an effective means of collaborating with others to achieve this goal
- they meet most of the characteristics of the general user[4]

The shared common interest, skill, or responsibility is the central component of the CoPE organization. There are any number of reasons why a set of individuals might want to collaborate on project involving the creation of a document. This characteristic alone is not sufficient. The shared commonality described above ads a crucial second component, beyond the first-order (immediate) need to collaborate – an incentive to engage in ongoing collaboration with the understanding that the result of such collaboration will product greater benefit than the sum of the work product itself.

The second characteristic follows relatively obviously from the first. The enhancement of the shared common interest, skill, or responsibility is a clear predicate for the expectation that collaboration will product a greater benefit than the sum of the work product itself. The third characteristic follows from the first in a similar fashion - if an effective means of collaboration already exists, the first condition will lack its predicate incentives.

---

[4] It is not necessary for a successful CoPE that *all* members be general (as opposed to more technically skilled) users. It is necessary that all (or nearly all) members at least meet the basic criteria of the general user.

We impose the fourth characteristic to conform to the type of groups which we seek to enable the formation of and within which we seek to enable collaboration.

3.2 User Interface Design

The requirements above indicated that a friendly, easy-to-use interface was paramount. This is particularly true for individuals/groups characteristic of the CoPE User Community.

To this end, the basic template for the CoPE was designed following the web-based column/widget model of organizing collaborative workspaces, which has become quite popular in this context. We developed a three-column interface for CoPE, comprising a primary workspace column (center), which displays the current focus document; a navigation column (left), containing the primary tools for navigating the CoPE hierarchy; and, finally, a discussion column (right) to provide easy access to common workflow tasks and related information to facilitate ease-of-use of discussion. The votable items window appears at the bottom of the discussion column. There is also a standard header/control bar across the top (see Figure 1).

An essential element of the design was that it functioned on a single page, requiring neither horizontal scrolling of the web browser nor multiple browser windows. This was primarily a usability decision; our participatory requirement process revealed that either change introduced substantial confusion to the General User. Surprisingly, however, vertical scrolling of the web browser did not. This was true provided that most/all of the users' control buttons/links were visible from the initial browser view (top of page). Interestingly, users experienced less difficulty when text entry fields (e.g., for Comments) extended below the viewing area. Almost no difficulty was demonstrated when the display of a document extended below the viewing area.

Finally, it was crucial that both the navigation pane and the top header/control bars were fully visible at first viewing of any page. It is also important online documentation was always visible in the navigation pane. To this end, complete and detailed, easy-to-read user manuals for the different types of existing users (e.g., general members and coordinators) were made available in CoPE. We found this feature tremendously helpful to users in familiarizing themselves with the system and feeling comfortable when moving to a new feature/page of the system.

3.3 Workflows

3.3.1 Overview

As discussed earlier, CoPE is a web-based collaboration system that is centered on documents. The CoPE system is designed to support the creation of documents, sharing them with the group, discussing a document, and ultimately making decisions on it as a group. This process is captured through a workflow where the document goes through different states as it progresses. Workflows can be customized to a given CoPE Community's needs through selection on discrete parameters. A CoPE coordinator organizes the activity of each CoPE site.

The following sub-sections provide a "walkthrough" of the workflow process. While relevant system design details are referenced throughout the description, it is assumed that the reader is at least familiar with the previous section on 3.2.

3.3.1.1 Creation of a Document. The first step in any workflow is the creation of a document. A document could contain any kind of information that the group might find useful and make decisions on. It could be a proposal, meeting minutes, or an article. This is the primary, essential task around which the entire CoPE system is designed.

When users log on to CoPE, they will see a screen like *Fig. 1*. On the left side of the screen is the navigation tree. The navigation tree is a visual representation of the structure and organization of the portal. The portal's content is organized in a hierarchical folder structure similar to what is found in most commonly-used computer operating systems.

The first step in creating a document is to find the folder in which the document should reside. When a folder has been selected, a window appears that allows the user to create a document or subfolder. A document is created by either entering text in the provided textbox or by uploading a Microsoft Word or PDF file.

When a document is first created it is in a 'draft' state. This is the initial state of the document's workflow as shown in *Fig. 2*. When a document is in 'draft' state it is only visible to the creator of the document and the coordinators of the CoPE.[5]

3.3.1.2 Sharing a Document. When a document is first created it is in 'draft' state. In this state, the document is only visible to the creator and not visible to the group.[6] The 'draft' state allows the creator to further edit and refine the document without it being observed by the group. When the user feels that their document is complete and ready to share, the user can make the document visible to the entire group by 'submitting' it. This is accomplished by clicking on the 'submit' action button, which moves the document in to a new state called the 'submitted'. This transition is detailed in the workflow diagram provided in *Fig. 2*. In this state, the document is visible to the entire group.[7]

---

[5] This section does not address the concept of system administrators. As there are a variety of means by which a CoPE may be deployed, it is possible (if unlikely) that a system administrator responsible for the "back-end" operation of the hardware, operating system, web server, and zope/plone platform upon which CoPE is based may not actually have access to draft documents.

[6] It is most likely however, as is the case in our Alpha Deployment, that a system administrator role will exist and will have full privileges to all objects throughout each CoPE on the system. The details of this role are not currently addressed in this document as future deployments are outside the scope of this writing.

[7] The concept of "subgroups", by which public visibility could be compartmentalized, is considered but not implemented in our research.

3.3.1.3 Discussion on a Document. Groups that have the ability to do so will often hold physical meetings to discuss matters and make decisions. As discussed in the section 3.1.1, however, groups using CoPE generally lack this ability.

Discussion helps the group to make better decisions by the sharing of opinions, assessing the pros and cons of an idea, and by bringing up new information. In this respect, the CoPE workflow enables an essential aspect of collaboration for geographically and/or temporally disparate groups. Similar to face-to-face interaction, discussions in CoPE are useful for group members to exchange opinions and ideas on a document. There are tradeoffs, however. CoPE discussions lack the realtime interactivity face-to-face meetings provide. Alternatively, CoPE compensates for this deficit in that it captures all of the discussion electronically for later review – something difficult to accomplish with face-to-face interaction.[8]

Once a document is in the 'submitted' state and shared with the group, users can engage in discussion by adding comments. Comments for a selected document are located on the right side of the screen in the comments tree. The comment tree can be seen in figure 1. A user can view the content of a comment by clicking on the title of the comment. The result is a "threaded-discussion" style of conversation widely adopted for newsgroups, message boards, blogs, and other similar Internet sites.[9]

3.3.1.4 Making a Decision on a Document. The goal of discussion is to produce some type of "actionable" outcome with regard to the document. We use "actionable" in a general sense; the outcome may be captured in the text of the document or it may literally be the approval of the document itself. CoPE supports this functionality by allowing the "approval" (or "disapproval") of documents through a voting procedure. The voting procedure is "democratic" in nature; CoPE permits a variety of voting schema based upon the principle of collaborative decisionmaking. For more information, see Section 3.3.2.

When a document is in 'submitted' state, the coordinator of the site can call a vote on the document. There are a variety of voting modes, the simplest just sets the number of positive votes required for the document to be accepted or 'approved'. When the document is being voted on, it is in the 'voting' state of the workflow. When a document receives the required amount of votes, it transitions from the 'voting' state to the 'approved' state. There are also facilities for archiving and reconsidering documents. The complete workflow for CoPE documents is shown in *Fig. 2*.

---

[8] Many of our test groups, including our own development team, often reflected that one of the greatest drawbacks of face-to-face meetings and/or conferences was the lack of codification of the discussion at such gatherings. For example, one member of our research team observed at a subsequent conference in a regularly scheduled series, attendees remarked "if we only remembered what we knew at the end of [the last conference], we'd be so much better off now." (This quote has been redacted for the protection of the subject(s) of the quote.)

[9] Relevant examples include: Slashdot (www.slashdot.org); various news sites commenting features (http://news.aol.com; [other examples?]); and online discussion groups (http://groups.yahoo.com). See also http://www.dartmouth.edu/~webteach/articles/discussion.html for a discussion of general principles of threaded discussion.

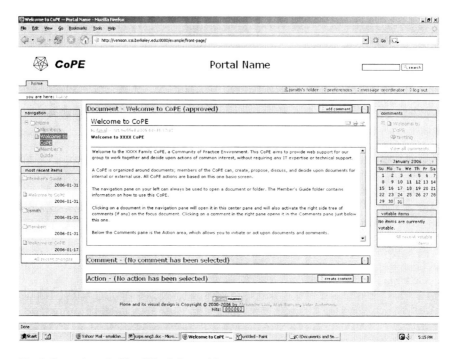

Fig. 1. Screenshot of a User Who Is Logged In

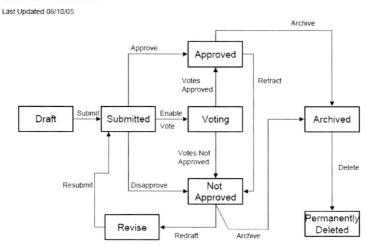

Fig. 2. CoPE Workflow for Documents. 'Draft' is the initial state.

3.3.1.5 Customization. Different groups have different organizational models and business processes. Also, their way of doing work may change over time. CoPE thus anticipates the possibility of selecting from multiple workflows. The default workflow shown here is a baseline we developed for archetype-style CoPE Communities from our participatory design process.

To accommodate for work process differences between groups, CoPE identifies a set of characteristics capable of being customized by a reasonably small set of parameters. By selecting on these parameters, an individual organization can adapt the environment to their workflow. We anticipate developing a first full set of parameters as we complete our Alpha Deployment testing.

3.3.2 Workflows and DCDCP

Workflows are the key to enabling DCDCP as described in Section 2. As discussed above, the workflow process is what enables a CoPE Community to create, disseminate, discuss, and act upon documents. It is important to recall here that documents are to be understood broadly – they are the vessel through which collaboration is enabled, not merely an end unto themselves. While CoPE can support the literal process of creating and approving documents, its true contribution comes in how the workflow process allows those documents to represent other actions and decisions facing the community.

3.4 Implementation Issues

Our review of the state of the art revealed that there was not then an obvious platform to support our system requirements. We experimented briefly with a hosted collaboration portal to get our first test group up and running and help develop system requirements. We quickly found that the system was too rigid in its design and too focused on commercial users to meet our requirements. Based on this analysis, we elected to pursue developing our own platform.

We began development of the CoPE system as an extension of the popular open-source web application platform Zope/Plone platform [11], [12].

Open-source projects that try to use the web for informal collaboration exhibit two characteristics that complicate their development. First, they lack the backing of a large, enterprise-scale hardware, operating, software, and security platform. Second, they require extended development times to reach even a first-iteration proof-of-concept since the work is, by nature, done in individuals' "free" time. These characteristics further complicate development by forcing developers to divide their efforts between advancing the project and maintaining proper security within their application.

Furthermore, the physical infrastructure of the server and the network on which it resides must also be considered with respect to security vulnerability when opening access to CoPE to the public Internet. Many of the Internet-based electronic attacks are the result of flaws discovered by malicious users [13]. In theory, a security vulnerability in CoPE's modification of Zope/Plone (or a similar vulnerability in Zope/Plone itself) could effect the application layers and result in a malicious user being able to execute code on the machine with privileged access.

4   Case Studies Supported by CoPE

We implemented the first "CoPE" as a single Plone instance on our Zope application platform. Since we were attempting to develop a platform that would meet the broad characteristics of requirements defined in sub-Section 3.1., multiple test groups were required.

To this end, we sought out groups that fit most of the CoPE User Community characteristics. These included an outreach program for students from local school systems, a small first-year section at a law school, an academic conference, and the faculty of multidisciplinary (both technical and non-technical) department at a major university. Our results from working with these groups were quite promising. Each represented a slightly different subset of each of the CoPE User Community characteristics. Since no "perfect match" groups were available, these more readily accessible groups were able to provide us with feedback and opportunities for iterative development with almost all the design elements in our requirements and specification.

The group with local school children comprised working with a program manager for the group and the students in the program. It was a natural fit to the CoPE Community coordinator/user roles. The program manager was the CoPE coordinator, and the students (and sometimes their parents) were the general users. While perhaps more technically sophisticated than some other users, the group was excellent for the "lack of preexisting collaboration utility" and "willingness to learn" characteristics of the general user.

The first-year law section was an interesting experience. As individuals, they possessed all the characters of General Users except one – the willingness to learn a new information/computing skill set. This lack of willingness was a function of the volume of work required of first-year law students; they simply lacked to time to learn how to use the system effectively enough to receive substantial benefit from it. The primary learning result of this experience was that willingness to learn was an absolutely critical factor that could block an otherwise well-matched group from being able to use the system.

The academic conference provided our first real test of a fully (geographically) disparate group. With the exception of the conference itself, none of the participants met face-to-face. The experience was surprisingly successful. In part, this may have been due to the conference coordinators' influence on directing participants to use the system prior to the conference. It may also, however, have been slightly biased as many (if not all) of the participants possessed a higher degree of technical skill than would a general user.

A similar logic applied to the academic department. The coordinator was the acting Chair of the department, and leveraged that position to encourage the members of the department to use the system. Again present was the secondary factor that many members of the group may have possessed higher technical skill. The heterogeneity of the group was an interesting aspect, however the testing period did not reveal any obvious insights into designing for such groups. This is an area that may warrant further inquiry.

5 Development Sustainability Issues

In the process of developing CoPE, we also learned a good deal about the problems with undertaking such a project in a research setting. Unlike with the private sector development of applications, we could not sustain a development and technical support team which could keep up with regular updates to core systems. In 2003, when our research began, this did not appear a heavily problematic challenge. However, as we continued our research, we developed several interesting – and unintentional – findings about the challenges of developing systems based on open-source platforms.

Projects of this nature exhibit two characteristics that inhibit their development sustainability. First, they lack the backing of a large, enterprise-scale hardware, operating, software, and security platform. Second, they require extended development times to reach even a first-iteration proof-of-concept since the work is, by nature, done in individuals' "free" time. These characteristics inhibit development by forcing developers to divide their efforts between advancing the project and maintaining proper security within their application.

Internet-based cyber-attacks continue to be a growing problem. Many of these attacks are the result of flaws discovered by malicious users. Most often, such "flaws" were not software "bugs" in the tradition sense of the word, but rather unanticipated vulnerabilities enabling attackers to compromise the application. As these vulnerabilities are discovered – both by successful attacks and by the application developers themselves – developers are forced to release incremental software updates, often referred to as "patches." Additionally, larger cumulative updates often referred to as "service packs" or "service releases" are generally released on a less frequent basis.

Both patches and services packs may involve substantial alterations to an application's codebase including changes that may render customizations based on earlier releases inoperable. Due to the responsive nature of patches (and services packs inasmuch as they incorporate or are due to discovered vulnerabilities), codebase changes that interfere with backwards compatibility may be unavoidable. More specifically, though the developers may wish to leave certain aspects of an application's API intact, discovered vulnerabilities in the implementation of that API may require that they modify, restrict, or even eliminate the offending portion of the API in order to close the security vulnerability.

The "ongoing upgrades" problem is further compounded in that it also applies to the virtual environment, web server, and operating system upon which the application depends. In the case of CoPE, therefore, the ongoing maintenance of several application layers must be considered in addition to the zope/plone platform itself:

- the Python virtual environment, upon which zope (and it's add-in plone) runs
- the web hosting environment (in our case Apache) which facilitates zope serving web content

- the development environment (in our case GNU/gcc/g++) through which both Python and Apache are compiled
- the operating system (in our case Red Hat Enterprise Linux) upon which the compiled applications run

Additionally, the physical infrastructure of the server and the network on which it resides must also be considered with respect to security vulnerability when opening up access to CoPE to the public Internet. In theory, a sufficient security vulnerability in CoPE's modification of Zope/Plone (or a similar vulnerability in Zope/Plone itself) could trickle down the application layers and result in a malicious user being able to execute code on the machine with privileged access.

6 Applications of "Democratic" CSCW

In this section we first suggest the main opportunities offered by DCDCM systems for the CSCL domain. We discuss the need of technological support for asynchronous interaction in on-line collaborative learning. Then, we justify the suitability of certain features of CoPE in this domain and introduce a new project we have started to extend CoPE to specifically support collaborative learning and the analysis of student activity. Finally, we conclude with some thoughts on the applicability of democratic CSCW.

6.1 E-Learning and DCDCM Systems

It has been understood for decades that, while lectures can be pre-packaged, there is no substitute for the interaction among students and teaching assistants in discussion sections. This has always been a problem for e-Learning, particularly where barriers of space and time prevent direct meeting.

Essentially any e-Learning system for students not in face-to-face contact must include some means to facilitate communication and interaction among students and instructors about the material. This requirement is the e-Learning analog of class "engagement." While there any many ad hoc approaches to this task, a democratic CSCW platform such as CoPE seems ideally suited to provide a systematic mechanism for both student-student and student-instructor interactions.

There are several features of the implemented CoPE system that support e-Learning discussions. Most obviously, the facility for hierarchical threaded discussion of documents can serve as a core for group consideration of material of any kind. Through the hyperlink facility, this can include arbitrary additional material. One obvious paradigm is to have the instructor post a document for discussion and to also intervene in the ongoing dialog when appropriate.

The CoPE mechanisms also support the production of joint projects by subgroups of students. It is easy to set up subgroups so that the work of each group is kept private from the others, but is visible to the instructor. Of course, all of the interaction ability is also available to the subgroup. This potentially has an advantage over traditional methods of direct interaction in that the instructor has access to (much of) the process of the group's effort and that this is well-codified for later review and use.

More recently, there has been wide spread use of interactive voting in the classroom. The basic idea is simple - a focused challenging (binary) question is posed to the class as part of a presentation of new material. There are several interesting variations on this theme. It is often useful to have small groups of students discuss the issue before voting. One can also use Delphi like techniques with repeated discussion and voting. This kind of classroom voting has proven to be quite successful and there is even a small industry providing electronic support for these techniques [13]. Of course, the voting mechanisms of a system like CoPE are ideally positioned to extend classroom voting to e-Learning. All of the alternative approaches to this pedagogical technique have natural realizations in CoPE.

Although, they have not been stressed in this paper, there are also mechanisms in CoPE that allow the coordinator of a CoPE site to customize much of the form and content of the material without programming. There is a coordinator's interface (and manual) that provides a range of choices on discussion and voting methods. Although this has not yet been verified, we believe that some variant of this functionality would enable instructors without IT expertise to customize their e-Learning discussion environments.

6.2 Extending CoPE to the E-Learning Domain

We are currently working on an extension of CoPE to provide full support for both formal and informal learning groups of the CoPE User Community type by applying the democratic discussion and decision-making mechanisms of the CoPE system to existing e-Learning applications.

A fundamental requirement to sustain CSCL applications is the representation and analysis of group activity interaction to facilitate coaching and evaluation [5]. Interaction analysis relies on information captured from the actions performed by the participants during the collaborative process. To this end, fine-grained usage data and other complex information collected from the learners' interaction are provided to give immediate feedback about others' activities and about the collaboration in general [3].

In extending CoPE to e-Learning, therefore, a primary requirement is extensive management and provision of information and knowledge in terms of task performance, group functioning and scaffolding [14]. The ultimate goal is to enhance and improve group activity by constantly keeping CoPE users aware of what is going on in the system (e.g. others' contributions, new documents created, etc.), In addition, monitoring participants' performance in CoPE allows tutors to identify problems that participants may encounter during the assignments. These findings can then be used to provide both real-time and asynchronous support to students (i.e., help students who are not able to accomplish the tasks on their own).

6.2.1 CoLPE Results

The results of this experiment came from a study of forty-three graduate students who were required to use an e-Learning extension of CoPE (termed "CoLPE" or "Communities of Learning Practice Environment") for a two week class assignment outside the normal scope of their online university. The results of this experiment [15] showed meaningful participation amongst the students. The finding was exploratory, but suggests promise in this application of democratic CSCW to e-Learning. By further intersecting CoPE CSCW system and CSCL principles we expect to provide more opportunities to learning by discussion and collaborative learning in general.

6.3 Further Thoughts

As we said at the outset, Democratic CSCW and e-Learning have overlapping but distinct goals and operating environments. The ongoing efforts, such as the extension of CoPE to the e-Learning domain, to integrate ideas from both literatures promise to yield both new insights and resulting systems that improve e-Learning and hopefully CSCW in general.

The cumulative results of our research suggest several findings. The development of workflows, for example – a process of abstracting system concepts into plain language – provides substantial benefit to non-technical users. This is essential in enabling the democratic collaborative environment described in this chapter. This benefit, however, cannot simply be foisted upon any given set of users. We had several setbacks and even failures throughout our testing, mostly with groups which did not satisfy the CoPE criteria. This revealed that while the potential for benefit is substantial, the benefit will best (if not only) manifest when the system is engaged by users who have a need for the type of democratic collaboration the system is designed to facilitate.

References

1. Held, D.: Models of Democracy, 2nd edn. Stanford University Press, Stanford (1996)
2. Koschmann, T.: Paradigm shifts and instructional technology. In: Koschmann, T. (ed.) CSCL: Theory and Practice of an Emerging Paradigm, pp. 1–23. Lawrence Erlbaum Associates, Mahwah (1996)
3. Zumbach, J., Hillers, A., Reimann, P.: Supporting Distributed Problem-Based Learning: The Use of Feedback in Online Learning. In: Roberts, T. (ed.) Online Collaborative Learning: Theory and Practice, pp. 86–103 (2003)
4. Bentley, R., Appelt, W., Busbach, U., Hinrichs, E., Kerr, D., Sikkel, S., Trevor, J., Woetzel, G.: Basic Support for Cooperative Work on the World Wide Web. Int. Journal of Human-Computer Studies 46(6), 827–846 (1997)
5. Dillenbourg, P. (ed.): Collaborative Learning. Cognitive and Computa-tional Approaches, pp. 1–19. Elsevier Science Ltd., Amsterdam (1999)
6. McGrath, J.E.: Time, Interaction and Performance (TIP). A Theory of Groups. Small Group Research 22, 147–174 (1991)
7. Thaw, D.: Communities of Practice Environment (CoPE). University of California, Berkeley, School of Information Master's Thesis (May 2008)
8. Stahl, G.: Groupware goes to school: adapting BSCW to the classroom. International Journal of Computer Applications in Technology 19(3-4), 162–174 (2004)

9. Feldman, J., Lee, D., Thaw, D.: Communities of practice environment. In: Morgan, K., Brebbia, C.A., Spector, J.M. (eds.) The Internet Society II: Advances in Education, Commerce & Governance, WIT Press (2006)
10. Benkler, Y.: The Wealth of Networks: How Social Production Transforms Markets and Freedom. Yale University Press, New Haven (2006)
11. Zope (web page as of November 18, 2007), http://www.zope.org
12. Plone (web page as of November 18, 2007), http://www.plone.org
13. Allen, J., Christie, A., et al.: State of the Practice of Intrusion Detection Technologies. CERT (web page as of November, 18, 2007)
http://www.cert.org/archive/pdf/99tr028.pdf
14. Daradoumis, T., Martínez, A., Xhafa, F.: A Layered Framework for Evaluating Online Collaborative Learning Interactions. International Journal of Human-Computer Studies; Special Issue on Theoretical and Empirical Advances in Groupware Research. Academic Press, Elsevier Lt (2006)
15. Caballe, S., Feldman, J., Thaw, D., Li, J.: Communities of Learning Practice Environment (CoLPE): An Extension of CoPE to the Learning Domain – Evaluation and Results. International Computer Science Institute – AI Research Group Report (January 2008)

Critical Thinking as a Framework for Structuring Synchronous and Asynchronous Communication within Learning Design-Based E-Learning Systems

Maria Kordaki[1] and Thanasis Daradoumis[2]

[1] Department of Computer Engineering and Informatics, University of Patras, 26500, Rion Patras, Greece
[2] Department of Computer Science Open University of Catalonia, Rambla Poblenou 156, 08018 Barcelona, Spain
kordaki@cti.gr, adaradoumis@uoc.edu

Abstract. This paper describes a conceptual framework focusing on the role that the development of learners' core cognitive skills and critical thinking plays on the success of synchronous and asynchronous communication within learning design-based e-learning systems. Based on this framework, we propose the design of specific tools which can be used by both teachers and students for structuring synchronous and asynchronous communication. In particular, a Cognitive Skill-based Communication Wizard (CSC-Wizard) is proposed as a supporting tool for helping discussion participants formulate appropriate interventions that express their intentions more clearly and thus facilitate the development of their cognitive skills more adequately. The design of this CSC-Wizard is based on modern social and constructivist views of learning and dialogue modeling. The idea, the rationale, the architecture and the interface associated with the proposed CSC-Wizard is presented through implementing a specific example within LAMS and MOODLE systems; which are widely used web-based, open source environments that support learning design.

1 Introduction

Thinking is essential as a foundation of learning. Philosophy and psychology foster also thinking as a framework of learning. In fact, the end product of education can be envisioned as 'the inquiry mind' [1]. In a nutshell, the centrality of teaching and learning within a framework that emphasize learners' cognitive development needs no justification, if the goal of education is not just to prepare learners to provide 'the right answers' to pass their exams but to create rational, mature thinkers who will be able to acquire and to appropriately use knowledge in analyzing problems, searching for meaning and make thoughtful decisions [2].

E-learning has provided many benefits in education in terms of flexible opportunities to learn anytime and anywhere to communicate and collaborate virtually throughout the world. For teachers, e-learning is seen as having the potential to reach new student markets, facilitating the tracking of student progress and activities as well as providing opportunities for creating innovating learning environments using modern both; theories of learning as well as tools and resources [3]. There is a plethora of e-learning environments and tools to support online learning. These tools include: a) communication, such as chats, forums, bulletin boards, etc. b) content presentation c) learning organization, such as group formation, timetabling, etc. d) learning assessment and e) searching. First generation of e-learning environments such as WebCT, and BlackBoard provided teachers and learners with the opportunity to use all these tools in integration. However, these environments seemed to not obviously support innovative or diverse learning activities. In fact, these are strongly based around information transmission [4].

Contrariwise, the 'learning design' based e-learning environments seemed as promising and revolutionary contexts for the design of pedagogically sound e-learning events. 'Learning design' has been defined [5] as an application of a pedagogical model for a specific learning objective, target group, and a specific context or knowledge domain. An important part of this definition is that pedagogy is conceptually abstracted from context and content, so that excellent pedagogical models can be shared and reused across instructional contexts and subject domains. In fact, a 'learning design' is defined as the description of the teaching-learning process that takes place in a unit of learning (e.g., a course, a lesson or any other designed learning event) [6]. The key principle in learning design is that it represents the learning activities and the support activities that have to be performed by different persons (learners, teachers) in the context of a unit of learning [7].

The IMS Learning Design (LD) specification aims to represent the design of units of learning in a semantic, formal and machine interpretable way [8]. LD has been related with: a) the use of ontologies and semantic web principles & tools; b) the use of learning design patterns; c) the development of learning design authoring and content management systems, and d) the development of learning design players, including the issues how to use the integrated set of learning design tools in a variety of settings [7]. Despite the fact that the IMS LD specification brings many pedagogical benefits when compared with earlier open specifications for eLearning, it is not easy for teachers to understand and work with it [9]. In fact, in the context of LD -non technologically experts- learning designers and teachers have difficulties because: a) no guidance is provided as to the kinds of pedagogic structures that they can create, b) the authoring using LD is not a simple task and c) the underlying concepts of the LD modeling language are not the same concepts that a teacher uses to think about in planning educational activities. Thus, the role of teacher - in the context of LD - is reduced to the role of a practitioner who has to use 'learning designs' ready-made by expert learning designers. This role implies the use of traditional behavioristic perspective of learning where learner individual differences are not acknowledged [10].

Contrariwise, modern constructivist and social views of learning [11; 12] emphasize that teaching is closely related with both; design of appropriate activities for each specific group of students as well as appropriate monitoring and intervention by the teacher during the learning process. According to these modern views, learners are in the center of the learning process -that means- that the learning tasks and activities have to be designed taking into account their previous knowledge and idiosyncratic characteristics. To this end, teaching can not become a predefined activity performed by remote experts in learning design but an online modeling, decision making and mediation process performed by the teacher as a basic actor in the design of the learning process whose interventions are also necessary. Among various type of teacher interventions are: proposing appropriate suggestions, asking sound questions, providing constructive feedback, proposing alternative representations and tasks as well as focusing attention, developing a positive attitude towards the tasks, expressing their thinking etc. [13]. Taking all the above into account, it seems clear that teachers need both high level tools to understand LD and easy-to-use tools which are specialized for a particular pedagogic context. It is suggested [14] to represent pedagogical practice in an appropriate form that teachers can easily apply, adopt, adapt, and reuse. At this point, it is also worth noting that a typical teacher needs training for the formation of appropriate learning activities and lesson plans.

Various examples of e-learning environments close to the LD specification are reported such as: RELOAD [15], CopperAuthor [16], COSMOS [17], MOT+ editor [18] and ASK-LDT [19]. However, these are mainly intended for expert designers and not for teachers. In contrast, the learning design languages for teachers in the creation of pedagogically sound learning designs are currently in infancy. For example, learning design languages such as LDVS [20, 21], LDLite [14], 8LEM [22] (Verpoorten et al. 2006) Learning Nuggets [23] usually have no explicit syntax and semantics specified. A learning design authoring tool based on Activity theory has been also reported [24]. In addition, the design of a tool that supports the design of questions to support students' basic cognitive skills has been recently reported [25]. There are also some integrated systems that support the idea of 'learning design' such as; Alfanet [26], LAMS [27] and MOODLE [28]. COLLAGE also is a system close to IMS LD specification that is friendly for teachers to use and supports collaboration using design patters [29]. It is worth noting that, the type of editor that usually classroom teachers need should be similar to the authoring environment provided by LAMS. In fact, LAMS offers a set of predefined learning activities, shown in a comprehensible way for teachers that can be graphically dragged and dropped in order to establish a flow chart of sequence of activities. Nevertheless, it was commented [27] that there is absence of tools supporting broader ranges of collaborative tasks and also missing support for the concepts of group creation and monitoring. Furthermore, there is an absence of tools that could support teachers' attempts for 'learning design' by explicitly taking into account the development of learners' cognitive skills. In addition, a coherent and integrated framework supporting the design of tools that could support teachers in learning design -especially on the design of tools that support both; synchronous and asynchronous communication- and focus on the development of learners' cognitive structures and critical thinking have not yet been reported.

Taking all the above into account, we have designed an e-communication editor; namely, a Cognitive Skill-based Communication Wizard (CSC-Wizard) to support primary and secondary level education teachers in their attempts at learning design, specifically in intervening in synchronous and asynchronous discussions so that to encourage the development of core thinking skills and critical and creative thinking in learners. This editor was designed taking also into account theoretical considerations arising from modern social and constructivist theories of learning as well as dialogue modelling.

This paper is part of a wider work [30] aiming at the design and the implementation of a system that would be appropriate for teachers so that they can encourage their students to develop their cognitive skills. In the following section of this paper, the rationale of the design of the proposed e-communication editor is presented. Next, the architecture of this editor is described and an example is demonstrated within the context of well known open source e-learning environments that support Learning Design, namely; the LAMS and MOODLE environments. Finally, the advantages of the provision of the proposed e-communication editor are discussed and conclusions are drawn.

2 The Rationale

2.1 Thinking Dimensions as a Framework of 'Learning Design'

In this section, an attempt has been made to concentrate on essential points of thinking dimensions, presented on the framework formed by [31] and to propose the design of computer-based communication tools that support the development of core thinking skills as well as critical and creative thinking by the learners. The aforementioned framework has been reviewed by numerous researchers, experts, practitioners and scientific organizations and also revised several times so as to be as accurate and helpful as possible. This framework has been proposed to be fully reflected in the design of learning curricula as well as in real teaching practices for the learning of each learning subject. Five dimensions of thinking have been identified, namely: a) Core thinking skills, b) Thinking processes, c) Critical and creative thinking, d) metacognition and e) the relationship of content-area knowledge to thinking. These dimensions reflect the various domain of thinking but do not form taxonomy. Usually, learners use these dimensions simultaneously -that means- they use core thinking skills and processes to solve a problem of a subject domain in critical and creative ways at the same time monitoring themselves and taking control of their learning. Next, we present a brief description of core thinking skills as well as of critical and creative thinking followed by a proposal of how to best structure discussion. The integration of these thinking dimensions with a dialogue model offers a solid base for the development of our cognitive communication tool that can be used in any an e-learning context.

a) Core Thinking Skills: These skills are used in metacognitive reflection as well as in thinking processes which are performed in the acquisition and performance of knowledge of each content area by the learners. Needless to say, these core skills are also implied in critical and creative thinking. Core thinking skills (TS_i, i=1,...21) have been classified into eight categories (C_i, i=1,...8) and are briefly presented bellow:

C1. *Focusing skills.* Two skills are included: TS1) 'Defining problems' that means clarifying situations that are puzzling in some way, and TS2) 'Setting goals'. These skills can be used at any time during a task to clarify/ verify and also redefine one's efforts.

C2. *Information gathering skills.* Skills included: TS3) 'Observing' involving obtaining information using learners' one or more senses, and TS4) 'Formulating questions' implying the focus on important information and searching for clarification of essential issues through inquiry.

C3. *Remembering skills.* Here, fall the skills of: TS5) 'Encoding', that is the process of linking pieces of information to be stored in long-term memory, and TS6) 'Recalling' that implies the use of effective strategies to store information for easy retrieval.

C4. *Organizing skills.* Here are included the skills of: TS7) 'Comparing' that means finding similarities and differences between or among entities, TS8) 'Classifying' that is grouping entities into categories based on some of their attitudes, TS9) 'Ordering' that implies the establishment of a criterion and the use of it to put entities in order or hierarchy, and TS10) 'Representing' that means put information in such forms (visual, verbal, symbolic), so that relationships of its critical elements be demonstrated in a meaningful way.

C5. *Analyzing skills.* Skills included in this category: TS11) 'Identifying attributes and components' that implies the analysis and recognition of the parts that constitute an entity, TS12) 'Identifying relationships and patterns' that means articulation of interrelationships among entities and recognition of the repetition of a pattern, TS13) 'Identifying main ideas' that is finding the main message or line in reasoning, and TS14) 'Identifying errors' involving the ability of detection of errors in logic and calculation procedures. These skills are crucial in the development of critical thinking.

C6. *Generating skills.* Here, fall skills such as: TS15) 'Inferring' implying the ability to go beyond available information to identify what maybe true based on learners' previous knowledge and reasoning, TS16) 'Predicting' that is the skill of anticipation of the progress and outcomes of a situation, TS17) 'Elaborating' that is improving understanding by adding relevant information and explanations.

C7. *Integrating skills.* Two skills included in this category: TS18) 'Summarizing' that means the learners' ability for condensing, selecting and synthesizing a cohesive statement from the data analyzed, and TS19) 'Restructuring' that is the ability of restructuring existing knowledge by incorporating new information.

C8. *Evaluating skills.* Here as well, fall the skills of: TS20) 'Establishing criteria' that implies the ability of establishing standards for judging about the value or logic of statements from both; philosophical and psychological points of view, and TS21) 'Verifying' that means confirmation or proving a statement by using the criteria of evaluation established using the previously mentioned skill.

Mapping core thinking skills to 'learning design'. Considering the core skills mentioned in this section, our framework proposes a specific vocabulary for critical thinking consisting of a number of appropriate key-words (see Section 3). These key-words can be used as labels in the construction of structured forums and chats, in the formation of relative questions, and in design patterns of learning tasks. To support teachers and students to successfully use these tools, our framework provides good practices of use of this vocabulary in designing good communication as well as appropriate questions and examples.

c) Critical and Creative Thinking. Both concepts are referred to the quality of thinking. Critical thinking has been defined as "reasonable, reflective thinking that is focused on deciding what to believe or do" ([32], p. 54). Important dispositions and abilities of critical thinking have been reported [33]. Creativity can be thought as 'the ability to form new combinations of ideas to fulfill a need' ([34], p. 324). Creativity has been related with: intense desire and preparation, internal locus of evaluation, reframing of ideas and working at the edge rather than the center of one's capacity.

Core cognitive skills that participate in critical thinking have been also reported [35] such as: TS22) Separation between facts and opinions. This skill implies the learner's ability to separate their own personal opinions which are arbitrary and some times biased from some facts that can be confirmed using specific data. TS23) Implementation-Improvement. This skill implies the learner's ability to transfer the knowledge constructed -in previous stages- in similar/analogous cases. Making also improvements of the solution constructed. TS24) Knowledge organization. This skill means that the learner is capable to form some diagrammatic visual hierarchical organization of the knowledge constructed during the data analysis and data transcendence stages of the experiment at hand. TS25) Empathy. This means the learner's ability to make sense of the other people's feelings and emotions of the situation at hand. So he/she can take a distance from a situation and accept the individual differences referred to it. TS26) Reflection. Reflection has been described as the mental process of looking back over the completed experience and performance to asses, analyze, and make connections to convert experience into learning and to lead to new understandings and appreciations [36] (Boud, Keough & Walker, 1985). Few people are able to convert personal experience to transferable learning, principles and models through the experience alone.

Mapping critical and creating thinking to 'learning design'. Our framework uses the previously mentioned dispositions and abilities of critical thinking in the design of learning activity design-patterns as well as of structured communications in forums and chat rooms. Motivating tasks and tools that support monitoring and self-evaluation can also be designed to enhance critical and creative thinking. In this paper we show how the vocabulary of critical thinking can be used to facilitate participants' intervention in forums and chat rooms.

2.2 The Role and Importance of Structured Communication within E-Learning Systems

This section examines how learning and knowledge building can be facilitated by supporting the development of learners' core cognitive skills and critical thinking in the context of well-structured synchronous and asynchronous discussions in a virtual learning environment. To this end, a conceptual sociolinguistic framework is defined for modeling dialogue and understanding how learning evolves and how knowledge is constructed during the discussion process. One important issue to consider is the types of interaction that occur in a discussion, the intentions which are manifested and finally the knowledge produced. This approach aims at identifying the various types of interaction produced and examining how an interaction type is related to the learning that results from it. As a result, this framework allows the study of how knowledge is transformed and becomes common to all discussion members.

In particular, this section examines how the building and distribution of knowledge is manifested in the context of teacher-student and student-student interaction and how it can be studied in a virtual learning environment. This involves the definition of appropriate learning situations that encourage the development of core thinking skills as well as critical and creative thinking by the learners, and the distinction of two levels of participant interaction; the discourse and the action level.

At the discourse level, the essential element is the interaction among peers (participants need to interact with each other to plan an activity, distribute tasks, explain, clarify, give information and opinions, elicit information, evaluate and contribute to the resolution of problematic issues, and so on). At the action level, task objects (e.g., documents, graphics) are created and manipulated. This approach focuses more at the analysis of the discourse level by seeing discourse as a medium and means through which the building and distribution of cognition is effected.

The framework proposed in this paper to support this model is based on an integration of several models and methods: the Negotiation Linguistic Exchange Model [37]; a model of Discourse Contributions [38]; and, the types of learning actions underlying a participant turn [39]. The structure of a long interaction is constructed cooperatively by using the exchange as the basic unit for communicating knowledge. Following [37], three general exchange structure categories are considered: give-information exchange, elicit-information exchange and raise-an-issue exchange, which consist of different types of moves (interventions) [40] and describe a generic discourse goal. More specifically, the goal of the actor who initiates the give-information exchange is to inform his/her partners about a certain situation with the aim to change the partners' mental states. Informing includes moves that explain, give an opinion, describe or remind a situation in different ways. The actor goal of the second exchange is to elicit the partners' state of mind (knowledge, beliefs, attitude, desire or abilities) of a situation which the actor is not aware or certain about. The actor goal of the third exchange is to raise an issue (a problem or question) to be resolved by the participants, which causes to explore their state of mind (knowledge, beliefs, etc.).

According to [37], there is a move that constitutes the "obligatory move" of the exchange, since it either carries or indicates completion of the discourse goal for which the exchange is initiated. The obligatory move of each of the above exchanges is: the first move of the give-information exchange, the second move of the elicit-information exchange and the third move of the raise-an-issue exchange.

According to [38], each move is seen as a contribution to discourse. This means that in a cooperative conversation, contributions are regarded as collective acts performed by the participants working together, resulting in units of conversation - typically turns (moves/interventions) - that aim to make a success of the discourse they compose. Yet, not all moves contribute in the same way toward the successful completion of the exchange.

Some moves have a pure contributing function toward the realization of the obligatory move of the exchange. This is the case of the first move of the elicit-information exchange, as well as of the first and the second moves of the raise-an-issue exchange. In fact, without the presence of those moves, the obligatory move cannot be realized; thus, those moves really contribute toward the realization of the obligatory move. Consequently, it is stated that successful realization of the obligatory move conveys evidence of (initial) success of the exchange [38].

In contrast, the other moves have a rather supporting function (provide evidence of support) toward the definite completion of the obligatory move and consequently of the exchange. This is the case of the follow-up moves of the three exchanges. Supporting moves are optional, so they may not be realized. In such a case, they convey an implicit support toward the obligatory move, that is, toward the definitive completion of the exchange.

Based on the work of [39], [41], and [42], partners are involved in a process of realizing a number of learning actions which lead to the completion of the exchange goal. Each move type captures and controls the evolution of the learning action performed by a participant by setting the expectations of the type of learning actions which has to be realized next by the other participants so that the goal set by the initial move be accomplished.

Completion of an exchange expresses the mutual beliefs of all participants about the accomplishment of its discourse goal. Moreover, it implies the achievement of a certain degree of knowledge building and distribution among the different participants. This degree can be deduced and measured by exploring the core thinking skills as well as the critical and creative thinking skills proposed by this model. As explained in next sections, for each participant the model can measure the way he/she contributes toward the development of a specific thinking skill by looking at the types of moves (interventions) that the participants create. The model can also deduce the users' participation behavior (focusing, organizing, analyzing, evaluating, etc.), as well as the effectiveness and impact that each move has in the discourse and in the achievement of the current discourse goal.

In general, the three types of exchanges represent standard discourse structures for handling information and suggest a certain type of knowledge building, as a result of giving and eliciting information or working out a solution on an issue set up. These discursive structures enable the participants to take turns, share information, exchange views, monitor the work done and plan ahead. Most importantly, they provide a means to represent and operationalize the cognitive product at individual level, that is, the way the reasoning process is distributed over the participants as it is shared in a collaborative discourse.

Consequently, interaction analysis takes into account both the way the interaction is structured and the types of contributions which are explicitly defined and expressed. The analysis of these interactions yields very useful conclusions on aspects such as individual and group working, dynamics, performance and success, which allows the tutor to obtain a global account of the progress of the individual and group work and thus to identify possible conflicts and monitor the whole learning process much better.

A further innovation of this model is that it allows participants to end up an exchange which took several moves to conclude by "replaying" the main contributing move of the exchange. For instance, in a set-up-an-issue exchange, a solution move may not be sufficiently complete and thus has to be further elaborated, corrected or extended. To that end, another participant has the option to provide an amplify-solution move which completes the initial solution. In general, a "replay" move can be used to resume all the changes produced from the initial appearance of an exchange goal to be achieved to its final conclusion and acceptance by all participants. This can be useful both to reinforce the fact that the goal of the exchange has been completed successfully and to explicitly indicate the progress achieved in the participants' process of knowledge building (especially as regards the participant who provided the main contributing move of the exchange).

Finally, the system requires the participant to commit certain action to indicate s/he is following a conversation, such as improve, support, assent or reflect upon a contribution. The aim is both to provide reliable indicators to measure the participants' critical and creative thinking skills and to promote the discussion's dynamics by increasing the users' interaction with the system.

Next, we show how the ideas presented in this model are further codified and implemented into a specific proposal of a Cognitive Skill-based Communication Wizard.

3 The Proposed Architecture for a Cognitive Skill-Based Communication Wizard

Communication is usually supported within e-learning environments synchronously and asynchronously by using the features of chat rooms and forums correspondingly. Learners and teachers can take advantage of these features, in terms of allowing diverse communications, especially from their own space and time. Despite these advantages, these features usually are very generic and are not enriched in such a way as to provide specific support for the user (teacher/student) to design their interventions within communication settings so that to encourage learners' cognitive skills. To this end, our proposed CSC-Wizard aims to act as a scaffolding tool for the design of communication interventions that support the development of critical thinking and core thinking skills in learners. In fact, nine different groups of Communication labels (CL) dedicated for the design of twenty six types of participant communication-interventions are proposed. These CLs have the form of appropriate 'words' which can be selected by the user (teacher/student) to form appropriate interventions in forums and chat rooms. Each type of CL is assigned to each different core thinking skill mentioned in the previous section of the paper. For example, labels CLTS5 are dedicated to the development of the thinking skill TS5, and so on. In fact, for each thinking skill, at least two carefully designed labels have been designed for use by the users. The architecture of the CSC-Wizard is presented in Table 1, including the aforementioned basic thinking

Table 1. Examples of Communication-Labels that could be used for appropriate intervention to develop basic cognitive skills in learners

List of Basic Core Thinking Skills	Communication Labels (CLTSij, i=1...26, j=1, 2)
C1. *Focusing skills*	
TS1: 'Defining problems'	CLTS01: Identify/State a problem
TS2: 'Setting goals'.	CLTS02: Set/Propose a goal
C2. *Information gathering skills*	
TS3: 'Observing'	CLTS03: Observe, Focus
TS4: 'Formulating questions'	CLTS04: Form a question, Request
C3. *Remembering skills*	
TS5: 'Encoding'	CLTS05: Encode, Codify, Check
TS6: 'Recalling'	CLTS06: Recall, Retrieve, Define
C4. *Organizing skills*	
TS7: 'Comparing'	CLTS07: Compare, Contrast
TS8: 'Classifying'	CLTS08: Classify, Categorize, Qualify
TS9: 'Ordering'	CLTS09: Order, Arrange
TS10: 'Representing'	CLTS10: Represent visually, Represent symbolically
C5. *Analyzing skills*	
TS11: 'Identifying attributes & components'	CLTS11: Identify attributes/ components
TS12: 'Identifying relationships & patterns'	CLTS12: Identify relationships/ patterns
TS13: 'Identifying main ideas'	CLTS13: Identify main ideas, Suggest main issues
TS14: 'Identifying errors'	CLTS14: Correct, Rectify
C6. *Generating skills*	
TS15: 'Inferring'	CLTS15: Infer, Deduce, Reason
TS16: 'Predicting'	CLTS16: Predict, Estimate, Provide
TS17: 'Elaborating'	CETS17: Elaborate, Process
C7. *Integrating skills*	
TS18: 'Summarizing'	CLTS18: Summarize, Conclude, Moderate
TS19: 'Restructuring'	CLTS19: Restructure, Modify, Replay
C8. *Evaluating skills*	
TS20: 'Establishing criteria'	CLTS20: Establish criteria/metrics
TS21: 'Verifying'	CLTS21: Verify, Ascertain
C9. *Critical and creative thinking*	
TS22: Separation between facts and opinions	CLTS22: Differentiate facts/opinions
TS23: Implementation-Improvement	CLTS23: Implement, Improve
TS24: Knowledge organization.	CLTS24: Organize, Structure
TS25: Empathy	CLTS25: Support, Understand
TS26: Reflection	CLTS26: Reflect, Think over, Acknowledge

skills (column, 1) while the proposed Communication Labels (CLTSij, i=1…26, j=1, 2) for the formation of each type of specific user-intervention are also presented in this Table (column, 2).

4 Implementing an Example of the Proposed CSC-Wizard within Learning Design Based E-Learning Systems

LAMS (Learning Activity Management System; http://www.lamsfoundation.org/) is an open source tool for designing, managing and delivering online collaborative learning activities. When using LAMS, teachers gain access to a highly intuitive visual authoring environment for the creation of sequential learning activities. These activities may be individual tasks, small group work or whole class activities. LAMS is based on the belief that learning does not arise simply from interacting with content but from interacting with teachers and peers. LAMS allows teachers to both create and deliver sequential learning activities which involve groups of learners interacting within a structured set of collaborative environments - referred to as 'learning design'.

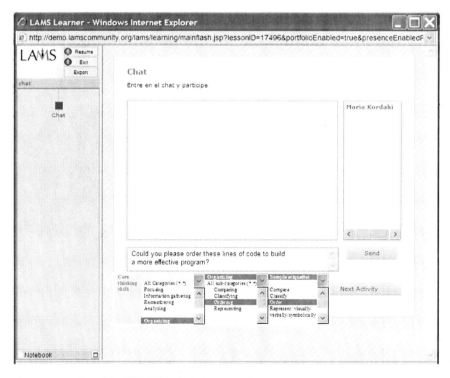

Fig. 1. Integration of the CSC-Wizard within a LAMS chat room

In essence, LAMS provides a practical way to describe multi-learner activity sequences and the tools required to support these. In fact, LAMS provides tools that support various activities such as presentation of information, writing and sharing resources, posing and answering questions as well as communication. MOODLE is also a similar Course Management System (http://moodle.org/). Despite this fact, the tools that support synchronous and asynchronous communication in both systems are very generic. Consequently, we suggest the integration of the proposed CSC-Wizard within the forum and chat room tools provided by LAMS and MOODLE (see Figures 1 and 2 for the CSC-Wizard integration within a LAMS chat and a MOODLE forum). As is shown in Figures 1 and 2, when the CSC-Wizard is integrated into the forum and chat room, participants are provided with the opportunity to construct the nine categories of interventions described in the previous section. They can select a specific intervention category, e.g. the intervention that supports 'Organizing' core thinking skills. Subsequently, this skill is presented with all its sub-skill types which are included in this category; namely 'Comparing', 'Classifying', 'Ordering' and 'Representing', (see the second pull down menu in Figures 1 and 2). At this point, a participant can select a specific type of Communication Label (e.g. 'order') that expresses the intention s/he wants to convey, and then proceeds to formulate an appropriate intervention that matches the label e.g., "Could you please order these lines of code to build a more effective program?" (see the third pull down menu and the question produced in Figure 1).

Fig. 2. Integration of the CSC-Wizard within a MOODLE forum

5 Summary and Future Research Plans

This paper has presented the idea and the architecture of an e-communication editor - the Cognitive Skill-based Communication Wizard, or CSC-Wizard - dedicated to supporting teachers and students in the realization of effective synchronous and asynchronous communication by forming such interventions that encourage the development of core thinking skills as well as creative and critical thinking in learners. The design of this editor has taken into account social and constructivist theories of learning as well as a sociolinguistic dialogue model. In fact, the CSC-Wizard consisted of twenty six – core, critical and creative thinking skills – labels that can be used to construct an equal number of types of communication interventions. These labels were designed to support the following core, critical and creative thinking skills: a) *Focusing*, including the specific cognitive skills of: Defining problems and Setting goals, b) *Information gathering*, including the skills of: Observing and Formulating questions, c) *Remembering*, including the skills of: Encoding and Recalling, d) *Organizing*, including the skills of: Comparing, Classifying, Ordering, and Representing, e) *Analyzing*, including the skills of identifying: attributes & components, relationships & patterns, main ideas as well as errors, f) *Generating*, including the skills of: Inferring, Predicting and Elaborating, g) *Integrating*, including the skills of: Summarizing and Restructuring, h) *Evaluating*, including the skills of: Establishing criteria and Verifying, and i) *Critical & creative thinking* including the skills of: Separation between facts and opinions, Implementation-Improvement, Knowledge organization, Empathy and Reflection.

Each type of these labels is dedicated to support the development of a core thinking skill. Integration of the proposed CSC-Wizard within the Forum and Chat rooms provided by MOODLE and LAMS is also presented. However, it is worth noting that the architecture of the proposed CSC-Wizard can be integrated into any e-learning environment that supports learning design. By using the CSC-Wizard, users have the opportunity to design communication interventions, not by chance but in a focused way, aiming towards the development of core, critical and creative thinking skills in learners. Having a solid theoretical base, the potential features of the proposed CSC-Wizard can find wide application in field studies which are deemed appropriate to test its impact on constructing real learning design by teachers.

References

1. Paul, R.W.: Critical thinking: Fundamental to education for a free society. Educational Leadership 42, 4–14 (1984)
2. Anderson, J.: The architecture of cognition. Havard University Press, Cambridge (1983)
3. Conole, G., Fill, K.: A learning design toolkit to create pedagogically effective learning activities. Journal of Interactive Media in Education 2005 (2008), http://jime.open.ac.uk/2005/08/

4. Britain, S., Liber. O.: A Framework for the Pedagogical Evaluation of eLearning Environment, VLEfullReport (2004), http://zope.cetis.ac.uk/members/pedagogy/files/4thMeet_framework/
5. Koper, R., Olivier, B.: Representing the Learning Design of Units of Learning. Educational Technology & Society 7(3), 97–111 (2004)
6. Koper, R., Tattersall, C. (eds.): Learning Design: A handbook on modeling and delivering networked education and training. Springer, Berlin (2005)
7. Koper, R.: Current Research in Learning Design. Educational Technology & Society 9(1), 13–22 (2006)
8. LD: IMS Learning Design. Information Model, Best Practice and Implementation Guide, Version 1.0 Final Specification IMS Global Learning Consortium Inc. (2003), http://www.imsglobal.org/learningdesign/
9. Griffiths, D., Blat, J.: The role of teachers in editing and authoring Units of Learning using IMS Learning Design. Advanced Technology for Learning 2(4) (2005), http://www.actapress.com/Content_Of_Journal.aspx?JournalID=63
10. Skinner, B.F.: The Technology of Teaching. Appleton, New York (1968)
11. Jonassen, D.H.: Designing constructivist learning environments. Instructional design theories and models 2, 215–239 (1999)
12. Vygotsky, L.: Mind in society. Harvard University Press, Cambridge (1974)
13. Wittrock, M.C.: Students thought processes. In: Wittrock, M. (ed.) Handbook of research on teaching, 3rd edn., Macmillan, NY (1986)
14. Oliver, R., Littlejohn, A.: Discovering and describing accessible and reusable practitioner-focused learning. In: Minshull, G., Mole, J. (eds.) Proceedings of Theme of the JISC Online Conference: Innovating e-Learning, pp. 30–33 (2006), http://www.jisc.ac.uk/elp_ conference06.html
15. RELOAD: Reusable eLEarning Object Authoring & Delivery project website (2005), http://www.reload.ac.uk/
16. OUNL: CopperAuthor project website (2005), http://www.copperauthor.org/
17. Miao, Y., Hoeksema, K., Hoppe, H.U., Harrer, A.: CSCL scripts: Modelling features and potential use. In: Koschmann, T., Suthers, D., Chan, T.W. (eds.) Proceedings of the Computer Supported Collaborative Learning 2005: The Next 10 Years!, pp. 423–432. Lawrence Erlbaum, Mahwah (2005)
18. Paquette, G., Léonard, M., Ludgren-Cayrol, K., Mihaila, S., Gareau, D.: Learning Design based on Graphical Modelling. Educational Technology & Society 9(1), 97–112 (2005)
19. Karampiperis, P., Sampson, D.: Designing learning services for open learning systems utilizing Learning Design. In: Uskov, V. (ed.) Proceedings of the 4th IASTED International Conference on Web-based Education, Grindelwald, Switzerland: ACTA Press, pp. 279–284. ACTA Press, Grindelwald (2005)
20. Agostinho, S., Harper, B., Oliver, R., Hedberg, J., Wills, S.: A visual learning design representation to facilitate dissemination and reuse of innovative pedagogical strategies in university teaching. In: Botturi, L., Stubbs, T. (eds.) Handbook of visual languages for instructional design: Theories and practices. Information Science Reference (2008)
21. Bennett, S., Agostinho, S., Lockyer, L., Harper, B., Lukasiak, J.: Supporting university teachers create pedagogically sound learning environments using learning designs and learning objects. International Journal on WWW/Internet 4(1), 16–26 (2006)

22. Verpoorten, D., Poumay, M., Leclercq, D.: The 8 Learning Events Model: a Pedagogic Conceptual Tool Supporting Diversification of Learning Methods. Interactive Learning Environments 15(2), 151–160 (2007)
23. Bailey, C., Zalfan, M.T., Davis, H.C., Fill, K., Conole, G.: Planning for Gold: Designing Pedagogically inspired Learning Nuggets. Educational Technology & Society 9(1), 113–122 (2006)
24. Miao, Y., van der Klink, M., Boon, J., Sleep, P., Koper, R.: Enabling Teachers to Develop Pedagogically Sound and Technically Executable Learning Designs (2008), http://dspace.ou.nl/bitstream/1820/1605/1/special_issue.miao.pdf
25. Kordaki, M., Papadakis, S., Hadzilacos, T.: Providing tools for the development of cognitive skills in the context of Learning Design-based e-learning environments. In: Bastiaens, T., Carliner, S. (eds.) Proc. of World Conference on E-Learning in Corporate, Government, Healthcare & Higher Education (E-Learn 2007), Quebec, Canada, USA, October, 15-19, 2007, pp. 1642–1649. AACE, Chesapeake (2007)
26. Van Rosmalen, P., Vogten, H., Van Es, R., Passier, H., Poelmans, P., Koper, R.: Authoring a full life cycle model in standards-based, adaptive e-learning. Educational Technology & Society 9(1), 72–83 (2005)
27. Dalziel, J.: Implementing Learning Design: The Learning Activity Management System (LAMS). In: Interact, Integrate, Impact, Proceedings ASCILITE 2003, Adelaide, December 7-10, 2003, pp. 593–596 (2003), http://www.ascilite.org.au/conferences/adelaide03/docs/
28. Dougiamas, M., Taylor, P.C.: Moodle: Using Learning Communities to Create an Open Source Course Management System. In: Dougiamas, M., Taylor, P.C. (eds.) EdMedia 2003, Honolulu, Hawaii, June 23-28 (2003)
29. Hernández-Leo, D., Villasclaras-Fernández, E.D., Asensio-Pérez, J.I., Dimitriadis, Y., Jorrín-Abellán, I.M., Ruiz-Requies, I., Rubia-Avi, B.: COLLAGE: A collaborative Learning Design editor based on patterns. Educational Technology & Society 9(1), 58–71 (2006)
30. Kordaki, M., Daradoumis, T.: Thinking dimensions as a foundation of learning design. In: Proceedings of the 9th IEEE International Conference on Advanced Learning Technologies, Riga, Latvia, July 14 -18 (2009) (to appear)
31. Marzano, J.R., Brandt, S.R., Hughes, C.-S., Jones, B.-F., Presseisen, Z.B., Rankin, C.S., Suhor, C.: Dimensions of Thinking: A Framework for Curriculum and Instruction. Association for Supervision and Curriculum Development, VA (1988)
32. Ennis, R.H.: Goals of critical thinking curriculum. In: Costa, A. (ed.) Developing minds: A resource book for teaching thinking, Association for Supervision and Curriculum Development, Alexandria (1985)
33. Ennis, R.H.: A taxonomy of critical thinking dispositions and abilities. In: Baron, J., Sternberg, R. (eds.) Teaching thinking skills: Theory and practice. W.H. Freeman, New York (1987)
34. Harpem, D.E.: Thought and knowledge. An introduction to critical thinking. Erlbaum, Hillsdale (1984)
35. Matsagouras, E.: Teaching Strategies. Gutenberg, Athens (1997)
36. Boud, D., Keogh, R., Walker, D.: Promoting reflection in learning: a model. In: Boud, D., Keogh, R., Walker, D. (eds.) Reflection: Turning experience into learning, pp. 18–40. Nichols, NY (1985)
37. Martin, J.R.: English Text: Systems and Structure. Benjamin Press, Amsterdam (1992)

38. Clark, H., Schaefer, E.: Contributing to discourse. Cognitive Science 13(2), 259–294 (1989)
39. Self, J.A.: Dormobilea: vehicle for metacognition. In: Chan, T.W., Self, J.A. (eds.) Emerging Computer Technologies in Education, pp. 1–20. AACE, Charlottesville (1994)
40. Schwartz, D.L.: The productive agency that drives collaborative learning. In: Dillenbourg, P. (ed.) Collaborative learning: Cognitive and computational approaches, pp. 197–219. Elsevier Science/Permagon, NY (1999)
41. Pilkington, R.M.: Analysing Educational Discourse: The DISCOUNT Scheme. Technical Report No. 99/2, Jan 1999, The University of Leeds, Leeds, England (1999)
42. Soller, A.: Supporting Social Interaction in an Intelligent Collaborative Learning System. Int. J. of Artificial Intelligence in Education 12, 40–62 (2001)

Constructing a Multi-agent System for Discovering the Meaning over Natural-Language Collaborative Conversations

Luis Casillas[1] and Thanasis Daradoumis[2]

[1] University of Guadalajara. Department of Computer Sciences. Av. Revolucion, 1500, 44840 Guadalajara. Mexico
luis.casillas@red.cucei.udg.mx
[2] Open University of Catalonia. Department of Information Sciences. Rambla Poblenou 156, 08018 Barcelona. Spain
adaradoumis@uoc.edu

Abstract. On the one hand, natural language is the main communication media for humans. It has a complex construction, based on the diversity of meaning for words and expressions according to the context. On the other hand, computers are not prepared to handle this ambiguity. Our work aims at presenting a multi-agent approach for dealing with the problem of discovering the meaning of expressions written in Spanish, based on a flexible recovery system and Bayesian principles. At a first stage, agents are supposed to identify the role of the words composing a sentence. At a second stage, a second set of agents is ready to coordinate among them in order to assemble a meaning. Our research forms part and contributes to the analysis of collaborative conversations among participants in a web-based collaborative learning environment.

1 Introduction

Natural language is the main communication media for humans. It has a complex construction, based on the diversity of meanings for words and expressions according to the context. Moreover, computers are not prepared to handle this ambiguity. Humans are capable of this kind of understanding due to common sense and reasoning; meanwhile computers are not able to perform that reasoning. Therefore, natural language processing (NLP) implies the construction of specific algorithms oriented to discover the meaning stored in certain sets of words. NLP is a field of the artificial intelligence, devoted to producing or understanding natural language.

The present work aims at presenting a multi-agent approach for dealing with the problem of discovering the meaning of expressions written in Spanish, based on a flexible recovery system and Bayesian principles. At a first stage, agents are supposed to identify the role of the words composing a sentence. At a second stage, a second set of agents is supposed to coordinate among them in order to assemble a meaning.

This mechanism will be used over de messages from discussion forums, in order to perform an analysis of conversations. The goal is to build a collaboration network based on the interaction acts detected by the conversation's analyzer.

2 Theoretical Bases

Artificial intelligence (AI) is mainly approached to problems and applications in which: there is no previous specifications for the problem or its solution, the specifications are not fully documented due to impossibility to drag the specific knowledge from humans, and the problem is so complex that no uniform specification could be done, the problem presents unexpected behaviors.

AI has introduced the concept of "knowledge based system", proposing a structure for dealing with knowledge in explicit, reviewable and accessible ways for users and programmers through the use of knowledge bases. Nowadays, such approach has enabled the construction of structured models, capable to deal with real problems. The knowledge engineering supports the construction of cognitive architectures, which include symbols and abstractions that represent knowledge and its management methods; providing and eventual intelligent-performance.

These trends in cognitive computing increased the abstraction requirements and complexity in systems. One of the ways to handle such complexity is the use of agents. An agent is a software entity capable of sensing its environment and act in order to reach its goals. Figure 1 sketches an agent in its environment. This paradigm allows a new way to develop solutions and to understand the AI. At same time, recent programming techniques provide the bases for the construction of advance solutions for complex problems. Modular approaches allow independent treatment to pieces; such as review, maintenance and reuse.

Besides, the intense use of the Web and other mobile devices demand new interfaces to search and manage huge volumes of information. Hence, a new interaction model is required. A framework capable to catch meaningful queries from users; oriented to the construction of cooperative and coherent dialogs among users and computers. It is a fact that an increasing number of people needs to interact with information systems. An advance interaction model must reduce the communication distance between the users and the machines, with a flexible, proactive and coherent approach. These whole ideas are oriented to the production of interfaces with NLP capabilities.

NLP is an ancient task in AI. The first stages of AI were devoted to manage the task of achieving in computers a behavioral-profile comparable to those shown by humans. This romantic-goal was set to reach an electronic brain, powerful as human's brain. It is known that AI is a field of computer's science devoted to the design and implementation of systems which try to emulate intelligent behaviors usually associated to humans.

In order to undertake such task, they are needed new techniques and paradigms which allow the representation and inference of knowledge; different strategies for the resolution of problems, NLP, learning procedures and other functionalities. Although there is a set of people with negative positions regarding computers' capacity to think, when computers become capable to perform real and deep chats

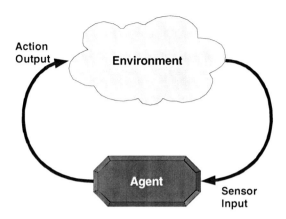

Fig. 1. An agent in its environment. Image adapted from [14].

with people; everyone will believe. Nevertheless the complexity of human's intellect versus the simplicity of the computer's capabilities, frequently led to dead-ends; which conducts to an impassable wall: handling the ambiguity present in the reality and manifested through the natural language.

Different efforts have been conducted in order to deal with NLP, most of them related with the English as object language. Science-fiction has presented diverse computers maintaining conversations with humans, although the computer H.A.L. 9000; shown in the movie "2001: a space odyssey" (directed by Stanley Kubrick and based on the novel of Arthur C. Clarke), is a remarkable example of an advance computer-treatment for NLP (and other high profiled behaviors for a machine!). In the reality, the most recognized project involving NLP is the classic ELIZA [14]; which is a clever system capable to support a conversation in an interactive fashion. The project ELIZA has inspired a significant number of interactive agents based on the Web. This kind of systems support simulations of conversations, but is common to arrive to dead-ends in which the conversation is blocked in empty argumentations.

Hence, it should be considered some kind of linguistic engineering; responsible for adapting the available knowledge resources to the application. The idea is to manage a set of primitive components of knowledge. Besides, different kinds of knowledge should be considered in order to perform NLP. Morphology knowledge is the most frequent technique, oriented to the extraction of roots; allowing the detection of similar words and avoiding the lost of significant meaning, and is highly recommended for flexible languages as Spanish. Lexical knowledge helps to deal the NLP through the perspective of grammatical categories: names, verbs, adjectives, and etcetera. Syntactic knowledge helps to identify the meaning in groups of lexical elements. Semantic knowledge allows the discovery of whole sentences based in the meaning of certain words or sets of words. Discourse knowledge is a highly specialized treatment which allows the detection of structures and organization used by authors. Pragmatic knowledge is oriented to the discovery of trends and intentions in the message.

A practical way to approach the NLP is through the analysis of word-roots (morphology). The Spanish language has a structure based in common roots, which direct to different words with slightly different meanings. Hence, in order to avoid machine confusions, all the discourse elements (words) will be analyzed in order to bind them under a common meaning. Thus, a set of agents will perform an analysis over the words forming the sentences of the discourse. This analysis is performed using retrieve-trees [1] for discovering common roots and the presence of the words in the vocabulary. The retrieve-trees for roots analysis were used successfully by [12] over Portuguese language, which is too similar to the Spanish. Besides, during the analysis we will use the Theorem of Bayes as in [5] to measure the probability for every word as member of specific word categories (article, substantive, adjective, verb and etcetera) through a lexical approach.

The agents have been also considered to support the efforts for handling the discovery of meaning from written expressions presented in natural language. The study [7] offers a framework for allowing users to instruct agents regarding the tasks to perform, this model is restricted for certain interaction forms to avoid ambiguity; unfortunately this approach restricts the communication to a query language model. The project [10] is also oriented to the development of specific interfaces for supporting the communication with users through agents devoted receive and process queries from the users. The solution proposed in [12] involves also the usage of agents, in this case the use of agents-technology is oriented to solve the problem of discovering morphology roles for every word, which is precisely the approach we are considering in our solution; nevertheless, the project shown in [12] has an unknown continuity.

The search for a semantic Web involves the NLP. The work [13] provides an approach for Web-search using complex expressions, based on the sub-products: NL-Page, a parser generator and MIN, a multi-agent framework. This machinery provides a powerful basement for supporting interactive agents during web searches, although our approach is not intended to the interaction between humans and agents, our agents will work over the messages written by humans when they interact; which is a most complex interaction model, away from our goals.

3 Conversations Analysis

This work has considered a specific approach regarding the discourse knowledge, implying structures and processes of knowledge settled in the joints between the inner thoughts-machinery and the language. A kind of interface between: cognitive process and linguistic activity. Apparently, the language is supported by a genetic need of humans to communicate and coordinate. Hence, language is used mainly to express communicative intentions inside interactive social structures. Language is, therefore, a mechanism to participate in conversations; the production and reception (exchange) of coordinated linguistic emissions involving one or more participants. Language production implies the existence communicating intentions in participants and certain plan to coordinate the emissions among participants; requiring for such tasks cognitive elements and dynamic response [11].

The use that everyone gives to language is related to his/her conduct in two ways: a) the capability of the person to handle grammatical aspects involving, the cultural competence; and b) the messages used to include contributions regarding the interaction goals, a pragmatic intention [11]. Hence, communication through a language is a complex activity; a demanding task involving planning and production under restrictions, some of them internal (the organization of participant's linguistic system) and some of them external (involving the communications context and conditions imposed by linguistic conducts).

Conversations might be understood as the prototype of discourse. Nevertheless, not every conversation implies a regular activity. Frequently conversations happen in a face-to-face fashion, filled of rich messaging in verbal and non-verbal elements. Some other times there are restricted scenarios, such as phone and radio conversations in which many non-verbal messages are lost (only language pausing and inflections prevail); in such scenarios the communication is made through a structured dialog that includes a strict control of turns in order to coordinate the messaging exchange. In these cases, language adopts sequential form that eventually acquires meaning and pragmatics in the context [3].

Another form of discourse is that in which there is not an immediate or direct response: talking to a public or writing an email, SMS or chat contribution. In such cases, the emission's receptor is not able to give an immediate response (due to time asynchrony or physical restrictions). Hence, the receptor will not contribute "on the air" to the conversation process; pushing to the message creator to produce his/her emissions over hypothetical bases regarding receptor's position. Emitter will make an additional effort to give the continuity for the discourse. Nevertheless this is considered a conversation, because there is a receptor to whose the message is oriented (even if receptor is passive or physically not present) [11]. We believe that this is, precisely, the kind of scenario manifested in the cooperative environments considered in our study.

From a cognitive approach, when the emission's receptor is not present (due to time or space); emitter's behavior is too similar to a monologue construction. The speaker must plan its messages, which become complex and complete in order to cover the absence of an active collaboration during the dialog [11]. Emitter should use linguistic and non-linguistic resources for supporting the coherence and interpretability in inner and external contexts.

At this point, any discourse could be understood as some kind of conversation; any conversation has a dependence function with the time and space, regarding the conversation elements provided by the other participants; and the monologues or dialog can be represented in oral or written expressions [9]. In conversations, the communicative intention is considered as the "emission component". Emissions supporting the communicative intentions used to be called "speech acts" [11].

Our system will deal with the messages leaved by participants during collaborative activity in a CSCL environment. The conversation analysis will be performed through the construction of collaboration networks among participants, assigning certain conversation roles to the bounds established between any pair of participants.

4 Agents Technology

An agent is a computing system settled in certain environment, as shown in figure 1, and is able to perform an action autonomously. Nevertheless, the agent has no full control over its environment; and in the best case has a partial control and limited sensoring.

Thus agents, as people, have limits to act and verify in their environments. There have been created a set of extensions for humans to perform the sensing and acting over their current environments: microscopes, telescopes, vehicles, computers, tools, etcetera. In the same way, agents might have extensions for its capacities for sensing and acting [14]. Nevertheless, in the case of humans there is a common-sense and agents running in computers will no have anything similar to common-sense; as explained before. Therefore, increasing agent capability to sense could

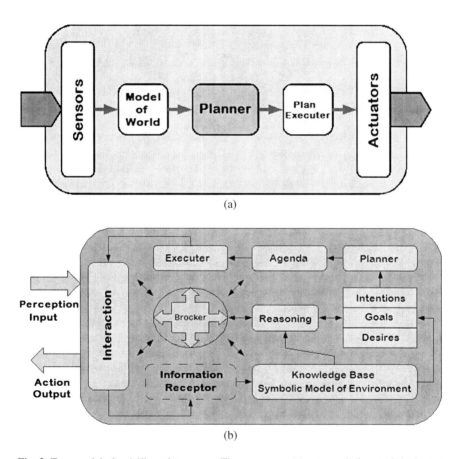

Fig. 2. Two models for deliberative agents. The upper one (a) a general view and the lower one (b) is a specialized approach. Adapted from ideas in [14].

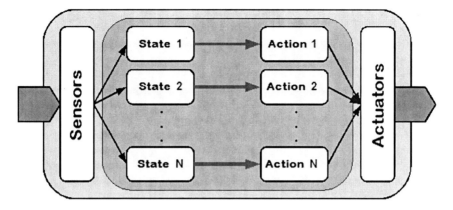

Fig. 3. Model of reactive agents. Adapted from the ideas in [14].

turn agents into chaotic entities; due to the absence of the common-sense to provide order.

The use of agents for NLP requires dealing with the complex task of meaning discovery inside discourses. A task usually performed by human's brains. Such situation invites to think in complex agents. Nonetheless, as mentioned in previous paragraph; increasing agent's capabilities could raise a set of negative situations. There are two main classes of agents, deliberative agents (shown in Figure 2) and reactive agents (shown in Figure 3).

A deliberative agent has special machinery inside, oriented to evaluate the context from a deeper approach and using inference engines involving historical elements. Meanwhile, reactive agents have a more pragmatic approach; they are based over the supposed that a plan is a set of actions. Of course, every action will be performed by the reactive agent and it will conduct to a new verifiable state. Reactive agents have limited collaboration, but they are able to collaborate indirectly.

For our research, they have been considered reactive agents, which give faster response and are easier to implement. Dealing with the huge volume of words in every grammatical category, demands a fast response; besides, we are not looking for discovering complex semantics from the messages considered during the conversation analysis.

5 Problem Description

A computer supported cooperative environment (CSCW: Computer Supported Cooperative Work, CSCL: Computer Supported Collaborative Learning) could include forums and chat-rooms. Our research is approached to the analysis of collaborations among participants of such environments, using discourse analysis as mentioned in previous section.

The first stage of such research was oriented to the quantitative analysis of the collaboration events, in order to produce a social network; which provides high level knowledge regarding the collaboration acts. The second stage, in which we are working on now, is devoted to the performance of a qualitative analysis of the conversations inside the forums and chat rooms. Such task involves NLP at certain level, as well as conversation analysis.

Our study must discover the intentions in the messages and the consequent reactions from the other participants in the same space (forum or chat-room), as well as the intentions in their respective reactions. Hence, an interaction network is built; now modelling the intentions in every direction.

NLP is a very specific task regarding the language involved. Although common-sense follows concrete rules (common to every human, regardless of his/her origins), we agree to think that every culture has developed a language machinery according to specific social development and depending on context and historical conditions; these facts imply different morphological analysis for every language and context in which it is used. Such behaviour depends, as it was stated previously, on human's genetic-machinery; oriented to support the communicative intentions.

Hence, the deepest need to communicate in humans (due to its gregarious nature) is translated to communicative intentions and the individual tries to express himself/herself through some available messaging structure. The messaging models are constructed as a historical accident, in which the environmental conditions and the current needs in the social group imply a specific ontology. The ontology is transmitted to every member of the group by his/her parents (in most cases) and later by other members of the same group. Ontology provides a naming system and the rules to express; somehow it includes the linguistic agreements that allow the communication process.

As we wrote before, our task will be performed over the Spanish language. Hence, our NLP will be implemented over messages written in Spanish language and with the specific goal of produce a simplified representation, oriented to recover a synthetic semantics.

6 Solution Proposed

We are not looking for a deep discovery of meaning from the conversations. A synthetic meaning will be enough to discover the collaboration acts among participants of the conversation. In order to achieve this treatment over the analyzed expressions, we have modeled agent-based machinery. Figure 4 sketches our approach for meaning discovery. The Word-Agents (smiling faces in fig. 1) are supposed to identify words belonging to specific categories in the sentence. Word-Agents use retrieve-trees [1] for a fast data-recovery and for providing the skill of binding two or more words with common roots (a frequent phenomenon in Spanish), as sketched in Figure 5.

Word-agents are also prepared for producing a probability value for every word in each category. Such probability is calculated according to the Theorem of Bayes [5]. The probabilities for calculating the Bayesian model are based on the impressing study [2], which provides a rich data-bank of frequency of words in the

Constructing a Multi-agent System

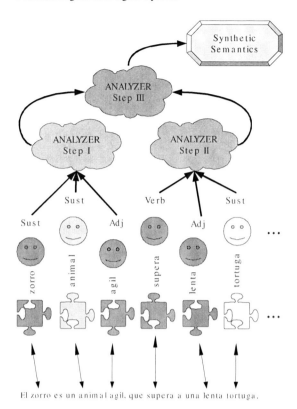

Fig. 4. Agent-based model for meaning discovery during NLP working over the Spanish sentence for: "The fox is an agile animal that passes a slow turtle".

Spanish language. These elements of information allow the Morphology-Agents (Clouds in fig. 4) to coordinate in order to produce the synthetic semantics following different steps.

The Bayesian principle (see formula 1) for learning is approached as a way to gather the certainty of the hypothesis related bind a word to certain category in the sentence. The Spanish word "bien" is an adverb and a substantive at the same time. Such word in Spanish is among the 175 most used, according to [2]. The average probability to use it as an adverb is 0.0013873025 and the probability for substantive is 0.0001328026. The probability to use that word as adverb is around ten times bigger to use it as substantive. These values are used to calculate the certainty of every word, for every category; the higher probabilities will imply the main meaning bound to the word.

$$P(B|A)=(P(A|B)*P(B))/P(A) \tag{1}$$

Morphology-Agents use their delegates (Word-Agents) to acquire a primary meaning for every word in the analyzed sentence, as well as the probability for being correctly settled as article, substantive, adjective, verb and etcetera; according to the responsibility assign every Word-Agent.

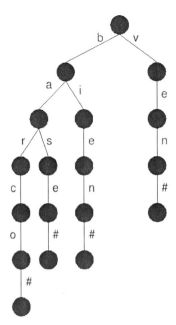

Fig. 5. Retrieve-Tree (Trie-Tree) storing the Spanish-words: "barco", "base", "bien" and "ven". Adapted from [1].

7 System Implementation

In order to build the structure of the system that will support the presented approach, it has been assembled a set of elements from the distributed systems' view. On the one hand, a middle layer is set to support the queries and results treatment among the applications of the system; on the other hand, a set of agents is raised to handle every category of word. As shown in figure 6.

The middle layer and the agents have been programmed in C# and tested over Windows and Linux (Ubuntu with Mono). The system includes both approaches for transferring packages over the network, TCP/IP Sockets and UDP Sockets. Datagrams (UDP approach) are oriented to achieving the construction of the network of agents, every agents report its IP address and its IP Port; the datagram is release and the middle layer collects all datagrams coming from agents. Later on, TCP/IP protocol is used to create persistent bounds between every agent and the middle layer, which improves performance by decreasing the transactions oriented to the connection.

Hence when an application is demanding attention from agents, it consults the middle layer (which stores the network location for agents) and the middle layer provides the reference to a thread that handles the socket bound to the specific agent involve in the current query.

Agents are built over a simple structure of objects. There is an object for representing words, a trie-tree object, an object for loading the words, an object representing the agent, two agents for sockets (TCP/IP, UDP) and an object for handling the files. Figure 7 sketches the agent's inner structure.

Constructing a Multi-agent System

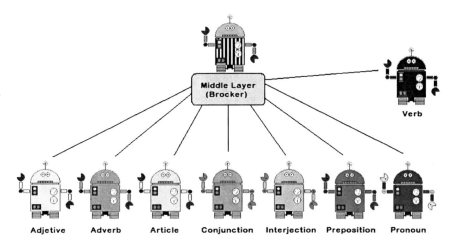

Fig. 6. Agents are referenced by a middle layer, which records their information regarding IP and IP-Port in which they are dispatching. The middle layer also works as an agent for the applications demanding responses from agents.

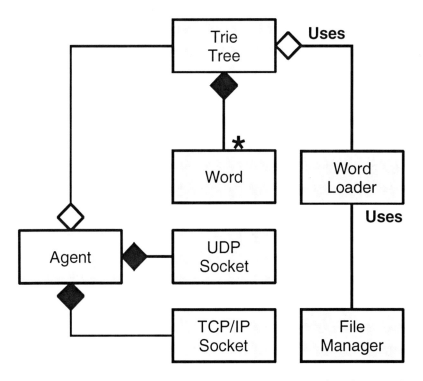

Fig. 7. Agents are instantiated based in this class diagram. There is an instance of the agent for every word category.

Agents load the words from different files. There is a file for every word-category. Every file contains the word id, word description, frequency, general probability and local probability (in the same category, based on its frequency over the total of words in the category). Table 1 shows some of the entries of one of these files.

The middle layer is the complement of this system. Its main work consists in storing the references for the agents. It is a kind of "yellow pages" book for the applications demanding the agents' response when trying to discover the role of words in a sentence. Hence, the middle layer has two main tasks: registering the agents when they arise, and attending the queries for agents' references. Thse last tasks are supported by the UDP socket. Figure 8 sketches the structure of the middle layer.

Table 1. Some of the entries from the file with the words for the category adverb, "adverbio"

897	(de)veras	Adverbio	36	0.000342	0.000021	remarkable
898	abajo	Adverbio	143	0.001359	0.000082	high
899	abiertamente	Adverbio	25	0.000238	0.000014	remarkable
900	absolutamente	Adverbio	137	0.001302	0.000078	high
901	acá	Adverbio	133	0.001264	0.000076	high
902	acaso	Adverbio	100	0.00095	0.000057	high
903	activamente	Adverbio	23	0.000219	0.000013	remarkable
904	actualmente	Adverbio	147	0.001397	0.000084	high
905	adecuadamente	Adverbio	23	0.000219	0.000013	remarkable
906	adelante	Adverbio	274	0.002603	0.000156	very_high
907	además	Adverbio	1019	0.009681	0.000581	very_high
908	adentro	Adverbio	69	0.000656	0.000039	high
909	adónde	Adverbio	40	0.00038	0.000023	remarkable
910	afortunadamente	Adverbio	32	0.000304	0.000018	remarkable
911	afuera	Adverbio	88	0.000836	0.00005	high
912	ahí	Adverbio	881	0.00837	0.000502	very_high
913	ahora	Adverbio	1950	0.018526	0.001111	very_high
915	allá	Adverbio	724	0.006879	0.000413	very_high
916	alrededor	Adverbio	170	0.001615	0.000097	very_high
917	altamente	Adverbio	23	0.000219	0.000013	remarkable
918	ampliamente	Adverbio	24	0.000228	0.000014	remarkable
919	anoche	Adverbio	59	0.000561	0.000034	high
920	anteriormente	Adverbio	49	0.000466	0.000028	remarkable
921	antes	Adverbio	1457	0.013843	0.00083	very_high
922	aparentemente	Adverbio	45	0.000428	0.000026	remarkable
923	aparte	Adverbio	80	0.00076	0.000046	high

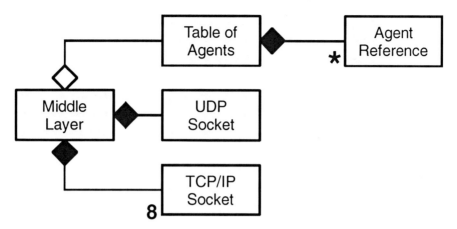

Fig. 8. The middle layer is settled over the agents and contains their IP references. Every agent could be in a different machine. Later on, when an application requires the services of one of the agents; the middle layer provides its IP reference through a thread containing the socket, thus the application can access directly and persistently to the agent required.

8 Conclusions and Future Work

NLP is a complex task. Our effort is not oriented to fulfil all the pending tasks in such field. We are developing a tool to handle a subset of the semantics underlying conversations in the spaces devoted to short communications among users in CSCW and CSCL environments. The current chapter is approached to order the work developed up to the moment in this task. Fully-functional algorithms (implemented in Mono-C# over Ubuntu and working in Windows) are currently working, in order to activate every piece of the whole model, and they have been developed the mechanisms to bind these pieces. When all the pieces are bound, we are enabled to process sentences; nevertheless it is not yet possible to deal with messages written by humans and discover the synthetic-semantics, although that is precisely our next step in this stage of our project.

At such moment, the compact meaning discovered will be compared to some reduced explanation for the very same messages; but these last will be process by humans, thus a qualitative comparison will allow the measuring of the success of the model.

References

1. Aho, A.V., Hopcroft, J.E., Ullman, J.: Data Structures and Algorithms. Addison-Wesley, Reading (1982)
2. Almela, R.: Frecuencias del español, diccionario y estudios lexicos y morfologicos. Universitas, Spain (2005)
3. Geurts, B., Beaver, D.: Discourse Representation Theory. Stanford Encyclopedia of Philosophy, Metaphysics Research Lab (2007)

4. Bianchi, D., Poggi, A.: Ontology based automatic speech recognition and generation for human-agent interaction. In: 13th IEEE International Workshops on Enabling Technologies: Infrastructure for Collaborative Enterprises, pp. 65–66 (2004)
5. Escolano, F., Cazorla, M.A., Galipienso, M.I., Colomina, O., Lozano, M.A.: Inteligencia Artificial: Modelos. In: Técnicas y Áreas de Aplicación. Thomson, Spain (2003)
6. Fum, D., Guida, G., Tasso, C.: A distributed multi-agent architecture for natural language processing. In: International Conference On Computational Linguistics; Proceedings of the 12th conference on Computational linguistics, Hungry, vol. 2, pp. 812–814 (1988)
7. Kemke, C.: An Architectural Framework for Natural Language Interfaces to Agent Systems. In: Proceedings of Computational Intelligence, San Francisco, California, USA (2006)
8. Keselj, V.: Multi-Agent Systems for Internet Information Retrieval using Natural Language Processing. Graduate Thesis, University of Waterloo. Canada (1998)
9. Mitchell, G.: Speech Acts. Stanford Encyclopedia of Philosophy, Metaphysics Research Lab (2007)
10. Rubin, S.H., Dai, W.: Natural Language Enabled Interface Agent. In: IEEE/WIC/ACM International Conference on Intelligent Agent Technology, USA, pp. 544–547 (2004)
11. Searle, J.R.: Speech Acts: An Essay in the Philosophy of Language. Cambridge University Press, Cambridge (1969)
12. Strube, V.L., Carneiro, P.R., Filho, I.S.M.: Distributing linguistic knowledge in a multi-agent natural language processing system: re-modelling the dictionary. In: Procesamiento del lenguaje natural, Spain, vol. (23), pp. 104–109 (1998) ISSN 1135-5948
13. Vlado, K.: Multi-agent Systems for Internet Internet Information Retrieval using Natural Language Processing. Ph.D. Thesis, Department of Computer Science; University of Waterloo, Ontario, Canada (1998)
14. Wooldridge, M.: An introduction to multiagent systems. John Wiley and Sons, Sussex (2002)
15. Weizenbaum, J.: ELIZA: A Computer Program For the Study of Natural Language Communication Between Man and Machine. Massachusetts Institute of Technology; Department of Electrical Engineering. Communications of the ACM 9(1), 35–36 (1966)
16. Yang, M.S., Yang, C.C., Chung, Y.M.: A natural language processing based Internet agent. In: IEEE International Conference on Systems, Man, and Cybernetics; Computational Cybernetics and Simulation, vol. 1, pp. 100–105 (1997)

CoLPE: Support for Communities of Learning Practice by the Effective Embedding of Information and Knowledge about Group Activity

Santi Caballe[1], Jerome Feldman[1,2], and David Thaw[1,2]

[1] International Computer Science Institute, 1947 Center St., Suite 600, Berkeley, CA 94704, USA
[2] University of California, Berkeley, School of Information,102 South Hall, Berkeley, CA 94720, USA
{scaballe, feldman, dbthaw}@icsi.berkeley.edu

Abstract. Communities of Learning Practice is an innovative paradigm focused on providing appropriate technological support to both formal and especially informal learning groups who are chiefly formed by non-technical people and who lack of the necessary resources to acquire such systems. Typically, students who are often separated by geography and/or time have the need to meet each other after classes in small study groups to carry out specific learning activities assigned during the formal learning process. However, the lack of suitable and available groupware applications makes it difficult for these groups of learners to collaborate and achieve their specific learning goals. In addition, they lack of advanced mechanisms of information management from the group activity for its further use in extracting and providing effective knowledge on interaction behavior. Indeed, this issue represents a fundamental requirement for current collaborative learning environments in order to adequately regulate the learning process as well as to enhance learning group's participation by means of providing appropriate awareness and feedback. In this chapter, we first describe the main guidelines that conducted the requirements and design of this. Then, we employ this tool in a real on-line learning environment to support a collaborative activity based on an asynchronous discussion. Finally, the experience and the evaluation results of using this application are reported, showing promising opportunities to support the formal and also informal discussion processes occurring in current communities of learning practice.

1 Introduction

Computer-Supported Collaborative Learning (CSCL) is an emerging paradigm dedicated to improving teaching and learning with the help of modern information and communication technology [1]. Its main goals are to create virtual environments where all the collaborative learning actors are able to cooperate with each other in order to accomplish a common learning goal. A fundamental requirement to sustain CSCL applications is the representation and analysis of group activity interaction to facilitate coaching and evaluation [2]. Interaction analysis relies on information captured from the actions performed by the participants during the collaborative process. To this end, fine-grained usage data and other complex information collected from the learners' interaction are provided to give immediate feedback about others' activities and about the collaboration in general [2].

Over the last years, collaborative learning needs have been evolving accordingly with more and more demanding pedagogical and technological requirements. To this end, on the one hand, modern pedagogical approaches targeting formal education include advanced learning techniques based on some form of collaborative consensus-building mechanism, such as learning by discussion and problem-based learning. Moreover, from the technological standpoint, a great deal of software packages in the form of Learning Management Systems (LMS) have recently appeared in the marketplace to enable the management of educational content and also integrate tools that support most of groupware needs, such as e-mail, discussion forums, chat, virtual classrooms, and so on. Representative LMS systems are Moodle and Sakai [3], which are being extensively adopted by educational organizations to help both educators create effective online learning communities, and educational institutions to highly customize the system to suit their pedagogical needs, and technological requirements.

On the other hand, informal collaborative learning is not receiving similar support from both research and technology standpoints and to the best of our knowledge this has been little investigated and exploited. One example is Yahoo groups, which provide support to virtual communities for free but they lack of essential groupware features, such as decision-making mechanisms and the provision of adequate feedback. Typically, students meet each other informally after classes in small groups to carry out specific learning activities assigned during the formal learning process. These groups of people form communities of learning practice where an important part of both individual and group learning process takes place and whose members are often separated geographically and have the need to meet asynchronously. However, the lack of appropriate support in the form of software tools makes it difficult for these informal groups to achieve their specific learning goals.

In this chapter, we present the main ideas in the form of a software platform to provide virtual communities of learning practice with advanced collaborative support especially designed to substitute the lack of pedagogical support to the learning process, such as the lack of a central knowledge authority and the support to students with poor technical skills. This system implements many of the approaches described so far and the first results drawn from real collaborative learning show very promising benefits for students in a real context of learning and in education in general. To this end, we present in Section 2 an existing system called CoPE developed by our research group that provides informal support to collaborative work and we also present the main guidelines of how to extend CoPE to the learning domain by incorporating essential functionalities according to the CSCL paradigm regarding the management of information and knowledge about group activity. Section 3 presents the development of the new platform that aims to provide students with new opportunities of learning outside of the formal education environment. The experience and the evaluation results of using this application in a real learning context are reported in Section 4. Finally, Section 5 will present the conclusions and ongoing work.

2 CoPE: Communities of Practice Environment

CoPE [4] is a web-based collaborative system aiming at providing formal and informal cooperative work over the Internet to non-technical people or those who lack of the necessary resources to acquire such systems. As such, CoPE provides most of the functionality expected from an asynchronous Computer-Supported Collaborative Work (CSCW) [5] application, such as information management and communication facilities.

CoPE is designed to enable a specific type of collaboration; a subset of CSCW that has not been adequately addressed so far. Specifically, this involves sets of individuals who share a need or desire to engage in collaborative production. The object of this production is something that can be codified in documents. CoPE is targeted to individuals who do not already have a formal workflow for this collaboration or who are seeking to improve upon inefficient workflows. CoPE also envisions enabling collaboration among individuals who are part of organizations with formal collaboration mechanisms, but whose mechanisms are limited to intra-organization collaboration. Finally, CoPE is designed to enable collaboration, not management, and thus envisions "democratic" collaboration.

There are many examples of sets of individuals around the world who have a need or desire to collaborate but lack the resources, knowledge, or institutions to do so. Consider, for example, public school teachers, social workers, and community action groups (where the group and its peer groups are the "individual"). Often these individuals are separated by geography and/or time. They could be too distant from one another to organize face-to-face meetings. They also could be unable to meet due to scheduling constraints or differing work hours. Such individuals may already be part of existing organizations but the "peers" with whom they wish to collaborate are in different organizations. CoPE is especially targeted to the individuals and organizations described here who lack substantial technical expertise or the resources to acquire such expertise. This includes any e-Learning situation for non-technical students.

CoPE is built by modifying and taking advantage of Plone/Zope's [6] powerful content management capabilities, such as information management, document workflow, and so on. CoPE modifies Plone appropriately to achieve the desired functionality.

2.1 Extending CoPE to the E-Learning Domain

It has been understood for decades that, while lectures can be pre-packaged, there is no substitute for the interaction among students and teaching assistants in discussion sections. This has always been a problem for e-Learning, particularly where barriers of space and time prevent direct meeting.

Essentially any e-Learning system for students not in face-to-face contact must include some means to facilitate communication and interaction among students and instructors about the material. This requirement is the e-Learning analogy of class "engagement." While there are many ad hoc approaches to this task, a democratic CSCW platform such as CoPE seems ideally suited to provide a systematic mechanism for both student-student and student-instructor interactions.

There are several features of the implemented CoPE system that support e-Learning discussions. Most obviously, the facility for hierarchical threaded discussion of documents can serve as a core for group consideration of material of any kind. Through the hyperlink facility, this can include arbitrary additional material. One obvious paradigm is to have the instructor post a document for discussion and to also intervene in the ongoing dialog when appropriate.

The CoPE mechanisms also support the production of joint projects by subgroups of students. It is easy to set up subgroups so that the work of each group is kept private from the others, but is visible to the instructor. Of course, all of the interaction ability is also available to the subgroup. This potentially has an advantage over traditional methods of direct interaction in that the instructor has access to (much of) the process of the group's effort and that this is well-codified for later review and use.

More recently, there has been wide spread use of interactive voting in the classroom. The basic idea is simple - a focused challenging (binary) question is posed to the class as part of a presentation of new material. There are several interesting variations on this theme. It is often useful to have small groups of students discuss the issue before voting. One can also use Delphi like techniques with repeated discussion and voting. This kind of classroom voting has proven to be quite successful and there is even a small industry providing electronic support for these techniques. Of course, the voting mechanisms of a system like CoPE are ideally positioned to extend classroom voting to e-Learning. All of the alternative approaches to this pedagogical technique have natural realizations in CoPE.

There are also mechanisms in CoPE that allow the coordinator of a CoPE site to customize much of the form and content of the material without programming. There is a coordinator's interface (and manual) that provides a range of choices on discussion and voting methods enabling instructors without IT expertise to customize their e-Learning discussion environments.

The extension of CoPE to e-Learning is called Communities of Learning Practice Environment (CoLPE), which will heavily rely on CoPE, and in turn on Plone, for most of the mentioned functionality that intersects CSCW and CSCL paradigms. However, specific behavior has to be aggregated to facilitate both the construction of knowledge among learners and the development of cognitive-acquisition skills, such as problem-solving abilities as well as the provision of an adequate multi-support framework so that tutors and peers can provide a suitable scaffolding when needed, as key aspects that distinguish CSCL from CSCW. CoLPE pursue theses objectives by means of seeing discussion as a medium through which the building and distribution of cognition is effected.

3 CoLPE Development

We are currently working on CoLPE to provide full support to both formal and informal learning groups by means of the collaborative discussion process. In this section, we present, first, the CSCL requirements that motivated the CoLPE development and, then, the main guidelines that conducted the design are described in certain detail.

3.1 General Requirements and Analysis

A fundamental requirement to sustain CSCL applications is the representation and analysis of group activity interaction to facilitate coaching and evaluation [2]. Interaction analysis relies on information captured from the actions performed by the participants during the collaborative process. To this end, fine-grained usage data and other complex information collected from the learners' interaction are provided to give immediate feedback about others' activities and about the collaboration in general [7]. To this end, in extending CoPE to e-Learning, therefore, a primary requirement is extensive management and provision of information and knowledge in terms of task performance, group functioning and scaffolding [8]. The ultimate goal is to enhance and improve group activity by constantly keeping users aware of what is going on in the system (e.g. others' contributions, new documents created, etc.). In addition, monitoring participants' performance allows tutors to identify problems that participants may encounter during the assignments. These findings can then be used to provide both real-time and asynchronous support to students (i.e., help students who are not able to accomplish the tasks on their own).

Learning by discussion forms an important social process where participants can think about the activity being performed, collaborate with each other through the exchange of ideas arising, propose new resolution mechanisms, and justify and refine their own contributions and thus acquire new knowledge. Aiming at these important objectives, CoLPE's requirements includes support to the essential types of generic contributions in a discussion process, namely specification, elaboration and consensus [8]. Specification occurs during the initial stage of the process carried out by the tutor or group coordinator who contributes by defining the group assignment and its objectives (i.e. statement of the problem) as well as how to structure the discussion (usually by discussion threads). Elaboration refers to the contributions of participants (mostly students) in which a proposal, idea or plan to reach a solution is presented by means of either contributing in an existing discussion thread or starting a new one. The other participants can elaborate on this proposal through different types of participation, such as questions, comments, explanations and agree/disagree statements. Finally, when a correct proposal of solution is achieved, the consensus contributions take part in its approval (this includes different consensus models, such as voting); when a solution arrives at consensus the discussion terminates.

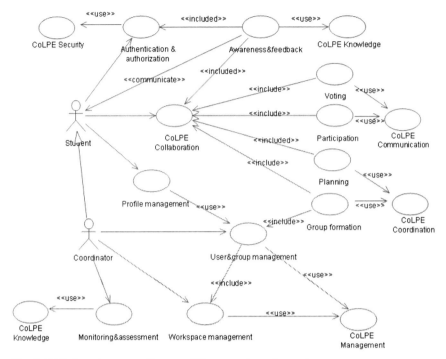

Fig. 1. UML-based use case diagram of the general requirements.

Other requirements are those considered common in most of domains, such as user management, security and system administration. User management refers to all logic related to the CSCL system user management as a group coordinator, group member, user-group and system administer as well as the user profile management. Authentication and authorization issues are fundamental needs in this context so as to protect the system resources from either the intentional or unintentional system use by the user as well as the system's access from unauthorized users. Finally, it is also essential in this context to carry out all system control and maintenance to correctly administer the system and aiming to improve in terms of performance and security. Fig. 1 summarizes and models formally all the requirements mentioned so far.

3.2 CoLPE Design

The CoLPE design aims at providing support to the essential types of generic contributions in a discussion process identified in the requirements, namely specification, elaboration and consensus. In CoLPE, these different types of generic contributions are managed by the three essential aspects existing in any CSCL application, namely coordination, collaboration and communication [8].

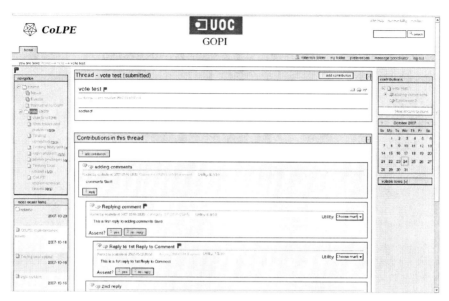

Fig. 2. Partial discussion view that shows several contributions in a thread. Each contribution bears information of its category the author chose before submitting it and peers' evaluation on average.

Specification phase is mainly based on coordination which involves the organization of groups such as workspace organization and group structure and planning, so as to accomplish group objectives. *Elaboration* phase is based on both collaboration and communication which allow students to share any kind of resources (e.g., participation spaces, documents, etc.) as well as exchange ideas by posting messages to a discussion space. During the elaboration phase, a key issue in CoLPE is that before a participant sends a new contribution to a discussion thread, this contribution is to be categorized by a predefined list of labels or categories, such as request for information, opinion, clarification, elaboration, etc.; inform in terms of extension, suggestion, explanation, justification, illustration, etc.; problem, which may be found as statement, solution, etc; greetings, motivation, among others (see [7] for more details). The purpose of these categories is to classify the intention of the contribution. Not all categories are always made available since depending on where the discussion is found just a subset of them are made available. These categories represent the information source to eventually present complex feedback to users in terms of participation impact and user profile (see further in this section for details). Finally, the *consensus* phase in the discussion process is also based on collaboration by which a voting system is shared by the group members to choose the best proposal.

During the discussion, participants may access different functionality available at contribution level (see also Figure 2):

- *Assent of contributions:* depending on the category of the contributions, certain contributions may be assented positively or negatively by the participant.

In case of a negative assenting, the participant is to explain the reason of this decision.
- *Reply of contributions:* always it is possible to reply a contribution by everyone (i.e., tutors and students). If there is no need for assent, then a chance to perform a normal reply will be provided.
- *Lecturer evaluation on participation quality:* lecturers are to evaluate the content quality of all contributions by reading them and assessing them. Please note this functionality is addressed to formal education only.
- *Peer evaluation on participation utility:* participants may assess others' contributions according to their usefulness in terms of the level of cognition achieved on the topic discussed moving forward in the discussion.
- *Reading contributions:* in order for a participant to make a read contribution be on record, it is not sufficient to visualize it. An explicit action has to be performed to show such an intention.

All user-resource and user-user interaction in CoLPE generates events or logs which are collected in log files and represent the information basis for the performance of statistical processes addressed at obtaining useful knowledge on the discussion process. This will make the collaborative learning process easier by keeping students aware of what is going on in their workspaces (e.g. others' contributions, the new documents created, etc.) as well as monitoring the general users' behaviour in order to provide appropriate support to them (e.g. helping students who are not able to accomplish their tasks on their own). In addition, this knowledge makes it possible to monitor and control the performance and general functioning of the discussion forum and hence it will enable the tutor to continuously track down the learning process and act if necessary. Finally, in order to efficiently communicate the knowledge achieved from the discussion process to students, will process and analyze the interaction data collected in a way that will provide full support to the presentation of this information in terms of awareness and feedback (see Figure 3).

- Feedback [7] goes one step further than awareness by providing exhaustive information of what is going on in the learning group over a long period of time (e.g. constantly showing to each discussion group member the absolute or relative amount of the contributions of others). CoLPE provides a multi-dimension model featuring activity, passivity, impact, affectivity, and assessment. These general indicators model each student's behaviour and performance and all of them are made available to all students and also tutors for monitoring purposes in both formal and informal context except for those who are applied to a specific context as mentioned (see also Figure 4):
- *Activity:* Participation behavior indicators are distinguished into proactive, reactive and supportive (or assentive).
- *Passivity:* Passive participants are considered those who just read others' contributions, as well as the ones who also evaluate the usefulness of these contributions.
- *Impact:* An impact value is assigned to users to measure the repercussion caused by their contributions to the discussion.

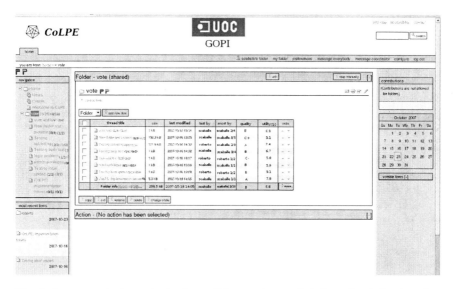

Fig. 3. Awareness is shown in the form of flags and numeric information informing about how many news, type, and location. Partial feedback at thread level is also provided informing of the quality and usefulness of the thread, among other indicators.

- *Effectiveness:* The effectiveness value of a move is calculated by the mean value of the number of assent moves received.
- *Assessment:* Tutor (in formal education) and peer (in both formal and informal education) assessment indicators are to evaluate both the quality of the contribution's content by the lecturer monitoring the discussion process and the usefulness of the contribution by the student participating in the discussion.

In order to equip CoLPE with appropriate knowledge management of the users' interaction data analysis, we took advantage of a generic, reusable service-oriented, component-based platform called Collaborative Learning Purpose Library (CLPL) [8], [9]. This platform enables a complete and effective reutilization of its generic components for the construction of specific collaborative learning applications. The CLPL is made up of five components in all, handling user management, security, administration, knowledge management and functionality (see [8] for a description of each component). The aim is both to map the essential elements involved in any collaborative learning application and provide specific support for interaction data analysis requirements as explained in Subsection 3.1.

To this end, this library is mainly performed by the two components, namely *CSCL Knowledge Management* and *CSCL Functionality*, which form the core of the CLPL in the construction of collaborative learning applications. They are briefly described here:

- The *CSCL Knowledge Management* component is made up of two subsystems, namely *CSCL Activity Management* and *CSCL Knowledge Processing* so as to support the first two stages of the information and knowledge management. The first subsystem manages the system log files made up of all the events occurring in a certain workspace over a given period of time. This event information is then classified according to a complete and tight hierarchy of events based on three general types of collaborative activity, namely task performance, group functioning and scaffolding [6]. The second subsystem performs the statistical analysis event information as well as the management and maintenance of the knowledge extracted by that analysis.

- The *CSCL Functionality* component implements the last stage of the information and knowledge management process, that is the presentation of the knowledge generated to users in terms of immediate awareness (see Fig. 1) and constant feedback (see Fig. 4) of what is going on in the system. In order to provide the essential awareness information to support collaboration, communication and coordination effectively, this subsystem defines three generic entities respectively, namely resource state, user status and group memory. Each of these abstractions acts as a vehicle so that awareness information can be classified and presented to users in the correct form depending on the type of activity involved. Finally, feedback information is achieved by defining certain generic entities such as history, pool and diagram and functions such as sorting. Based on these abstractions it is possible to dynamically gather and store a great amount of history data and statistical results from group activity. For the purposes of presentation format, this component defines a flag as a single abstraction supporting the presentation of awareness information to users through the user interface as well as a chart for the presentation of complex information in the appropriate diagrammatic format (e.g., pie chart, histograms, plain text, etc.).

Finally, personal features of the discussion group participants (their role, collaboration preferences and so on) were taken into account and a user and group model were designed so as to allow participants to add new services as their needs evolve as the discussion moves forward. All these user features were included by the CSCL User Management component through the user profile management subsystem, providing solid support for building and maintaining the user and group model.

User and group management is also provided by CoLPE mainly by means of CoPE's resources allowing tutors and group coordinators the management and maintenance of the user data and the learning groups. User and group profile is also to be set up by students to personalize the system according to their needs and preferences, such as personal information, position, language, connection status and so on. In the same way, a module is included to manage and set up the workspaces by assigning them the necessary resources (e.g. agenda, calendar, discussion place, voting system, etc.) so that learning activities can take place.

Finally, security issues will be managed by two main modules: user access (i.e. authentication) and privilege assignment (i.e. authorization) both aiming at restricting groups to accede to others' valuable resources and making tutors easier to track all groups. To this end, CoLPE will rely throughout on CoPE's authentication and authorization mechanisms.

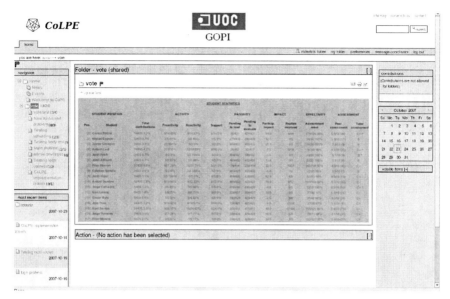

Fig. 4. Full feedback information presented to all students. Each student can compare his/her individual performance to the rest of the group.

To sum up, the ultimate objective of CoLPE is to fully provide functional support to the discussion process as part of the current pedagogical models in both inside and outside the educational space. Moreover, this application supports the mentioned process of embedding information and knowledge about group activity into the discussion process so as to provide the participants with immediate awareness and constant feedback of what is going on in the discussion.

4 Evaluation and Results

In order to evaluate our prototype of CoLPE and analyze its effects in the learning experience, and in particular the discussion process, we used the real on-line learning context of the Open University of Catalonia[1]. 43 graduate students enrolled in the course Methodology and Management of Computer Science Projects were involved in this experience.

4.1 Experiment Procedure

The experience consisted of a discussion assignment, with the aim of discussing how a project manager can deal with the problem of changing the requirements of

---

[1] The UOC is located in Barcelona, Spain, and offers full distance higher education to currently 50,000 students through the Internet since 1994. The virtual campus also supports about 2,000 lecturers and tutors who are involved in some of the 600 on-line courses available from 23 official degrees and other PhD and post-graduate programs. The UOC is found at http://www.uoc.edu

software projects which are already in advanced phases of their development because of demanding and urgent needs of the clients. The assignment title was: "Change management: necessity or virtue?".

The procedure was the following: students were free to as many discussion threads (i.e., head of threads) as wished to appropriately deal with the discussion topics. Any student could then contribute in both his own and any other discussion threads as well as he/she can start extra threads to provide new arguments or approaches with regards to the issue addressed. The only requirement was to make at least one post to either a head of thread or a comment.

4.2 Results and Analysis

The results of this experiment are provided by means of statistical analysis. A structured and qualitative report was also conducted at the end of the discussion addressed to all participants who were asked to both assess the prototype and compare it to the standard well-known discussion tool they had already used in previous courses at the UOC.

A statistical analysis of the results of the discussion is shown in Table 1. Note that the discussion took place at the end of the course and even though the number of potential participants was 43 (i.e., students enrolled in the course), roughly 40%[2] of them had already made the decision to give up before the assignment started and as a result they did not pay attention nor contribute to the discussion. So, the number of active participants who participated in the discussion actively or passively was 26.

Table 1. Basic statistics about participation

Statistics	CoLPE
Number of students enrolled	43
Number of students actually participating	26
Number of heads of thread	17
Number of comments in threads	93
Total posts	110
Mean number (posts /thread)	M=6,4 SD=4,5
Mean number (posts /student[3])	M=4,2 SD=3,8

From the results of Table 1, the SD statistic for the posts/thread mean appears to be high, which shows the heterogeneity of the discussion involving threads of very different length and also that actual discussion was generated and as a result the contributions became highly structured and specific. In addition, the posts/student mean rates high (the requirement was 1 post per student) and shows a general interest in the discussion.

---

[2] Currently, the drop-out average at the Open University of Catalonia is about 50%.
[3] Students who participated in the discussion.

On the other hand, the SD statistics for posts/student is also high meaning that some students participated a lot (more than 10 posts) while a few tried to fulfill the assignment's requirement and provided single, monolithic point of view. It could be argued that at the end of the course students lack time, though more experimentation has to be undertaken to confirm these results.

Table 2. Distribution of the tagged contributions

Exchange actions	Contribution categories	# Tagged contributions
support	Greeting	3
	Motivation	0
request	REQUEST-Information	1
	REQUEST -Elaboration	0
	REQUEST -Clarification	3
	REQUEST -Justification	0
	REQUEST -Opinion	20
	REQUEST -Illustration	0
	INFORM-Extend	17
inform	INFORM-Lead	0
	INFORM-Suggest	8
	INFORM-Elaboration	0
	INFORM-Explain/Clarification	17
	INFORM-Justify	1
	INFORM-State	0
	INFORM-Agree	21
	INFORM-Disagree	6
	PROBLEM-Statement	16
set-up-an-issue	PROBLEM-Solution	1
provide-solution	PROBLEM-Extend solution	0

Table 2 shows the most frequent categories used to tag the contributions. Although the choice of the category appears to be mostly correct, they could indeed be more precise. The permanent availability of all possible categories did not help participants to choose carefully. In future iterations, only those categories which are appropriate (i.e., make sense) at a certain point of the discussion will be shown, thus facilitating the choice a great deal.

Table 3 shows the results of a structured and qualitative report conducted at the end of the discussion addressed to the CoLPE users who were also asked to compare it to the standard well-known tool they had already used in previous courses at the UOC. This report shows the technical problems we faced due to the server where CoLPE was installed (Linux SuSE 2.4.21-99 machine, Intel Pentium 4 CPU 2.00 GHz, 256MB RAM) performed poorly and it was unable to conveniently handle both the demanding hardware requirements of Zope and the participants' concurrency.

Table 3. Excerpt of the questionnaire filled out by the students

Selected questions	Average of structured responses (0 – 5)	Excerpt of students' comments
Asses the CoLPE application	4	"Despite technical problems with the server I found CoLPE very useful due to the distribution of posts into threads and also be aware of where the news was" "I liked the categorization as it helped me understand others' contributions and reply being more confident on my contribution" "The notification of news was useful" "CoLPE is more suitable to support this type of discussion than the UOC's forum" "Certain functions are missing in CoLPE: subscription to your thread, advanced search function, …"
Evaluate how CoLPE fostered your active participation	2	
Did CoLPE help you acquire knowledge on the debate's issue?	2	
Compare CoLPE to the UOC campus' standard forum tool	3	

5 Conclusions and Future Work

This chapter describes an extension [10] of a promising approach for enhancing communities of learning practice by means of an innovative tool that contributes to the improvement of the discussion process occurring in both formal and informal collaborative learning settings. In addition, the provision of effective information and knowledge about group activity is a major concern in this contribution since it may enhance and improve the specific type of collaborative learning addressed in this chapter a great deal.

As such, the main ideas provided in this chapter are not conclusive due to its exploratory nature. To this end, we report the experience of using this prototype in a real context of on-line learning, though the results are not conclusive due to its exploratory nature. However, the analysis of the results promises significant benefits for students in the context of project-based learning, and in education in general.

More powerful hardware will be used in the next experiments so as to overcome the poor server performance issue. Moreover, a decentralized distributed infrastructure is intended to be added to the CoLPE prototype in order to meet other important non-functional requirements that may influence the learning process a great deal [11], such as scalability, fault-tolerance, and interoperability. For instance, the gain in fault-tolerance might help enhance the effectiveness of complex collaborative learning processes (e.g., by avoiding a central point of failure). We plan to explore these interesting possibilities in the next iterations of the CoLPE design.

Acknowledgments

This work has been partially supported by the Catalan Government (AGAUR) grant BE-1 10025.

References

1. Koschmann, T.: Paradigm shifts and instructional technology. In: Koschmann, T. (ed.) CSCL: Theory and Practice of an Emerging Paradigm, vol. (1-23). Lawrence Erlbaum Associates, Mahwah (1996)
2. Dillenbourg, P.: Introduction; What do you mean by "Collaborative Learning?". In: Dillenbourg, P. (ed.) Collaborative learning. Cognitive and computational approaches, pp. 1–19. Elsevier Science, Oxford (1999)
3. Caballé, S.: On the Advantages of Using Web & Grid Services for the Development of Collaborative Learning Management Systems. In: Proceedings of the 3PGIC 2007, Vienna, Austria (2007) ISBN: 0-7695-2823-6
4. Feldman, J., Lee, D., Thaw, D.: Communities of practice environment. In: Morgan, K., Brebbia, C.A., Spector, J.M. (eds.) The Internet Society II: Advances in Education Commerce & Governance. WIT Press (2006)
5. Bentley, R., Appelt, W., Busbach, U., Hinrichs, E., Kerr, D., Sikkel, S., Trevor, J., Woetzel, G.: Basic Support for Cooperative Work on the World Wide Web. Int. Journal of Human-Computer Studies 46(6), 827–846 (1997)
6. Latteier, A., Pelletier, M., McDonough, C., Sabaini, P.: The Zope Book. Zope.org. (2007), http://www.zope.org/Documentation/Books/ZopeBook
7. Zumbach, J., Hillers, A., Reimann, P.: Supporting Distributed Problem-Based Learning: The Use of Feedback in Online Learning. In: Roberts, T. (ed.) Online Collaborative Learning: Theory and Practice, pp. 86–103 (2003)
8. Caballé, S., Daradoumis, T., Xhafa, F.: A Generic Platform for the Systematic Construction of Knowledge-based Collaborative Learning Applications. In: Caballé, S. (ed.) Architecture Solutions for e-Learning Systems, vol. XII, pp. 219–242. Idea Group Inc. (IGI) (2007)
9. Caballé, S., Daradoumis, T., Xhafa, F.: A Model for the Efficient Representation and Management of Online Collaborative Learning Interactions. In: Cunningham, Cunningham, M. (eds.) Proceedings of e-Challenges 2006, Building the Knowledge Economy: Issues, Applications and Case Studies, Barcelona, Spain. IOS Press, Amsterdam (2006)
10. Caballé, S., Feldman, J., Thaw, D.: Supporting Communities of Learning Practice by the Effective Embedding of Information and Knowledge into Group Activity. In: Proceedings of CESA 2008, Barcelona, Spain. IEEE Computer Society, Los Alamitos (2008)
11. Caballé, S., Xhafa, F., Daradoumis, T.: A Service-Oriented Platform for the Enhancement and Effectiveness of the Collaborative Learning Process in Distributed Environments. In: Meersman, R., Tari, Z. (eds.) OTM 2007, Part II. LNCS, vol. 4804, pp. 1280–1287. Springer, Heidelberg (2007)

Detecting and Solving Negative Situations in Real CSCL Experiences with a Role-Based Interaction Analysis Approach

José Antonio Marcos-García[1], Alejandra Martínez-Monés[1], Yannis Dimitriadis[2], Rocío Anguita-Martínez[3], Inés Ruiz-Requies[3], and Bartolomé Rubia-Avi[3]

[1] School of Computer Science Engineering
[2] School of Telecommunications Engineering
[3] Faculty of Education
University of Valladolid, Camino del Cementerio s/n, 47011, Valladolid, Spain
{jamarcos, amartine}@infor.uva.es, yannis@yllera.tel.uva.es,
{rocioan, inesrure, brubia}@doe.uva.es

Abstract. Collaborative learning has a number of potential benefits, which do not always occur, partially due to the difficulties that students and teachers have to establish good social interaction patterns. These interaction patterns depend on the roles assumed by participants in the learning process. In real practice, teachers need support to be able to detect these emergent roles and undesired interaction patterns, especially if collaboration is mediated by computers, and thus is not directly observable by humans. Interaction analysis (IA) methods and tools are adequate to support the regulation of the collaborative activities, using the analysis results to provide adequate feedback to the different participants in their specific roles. We have proposed a role-based approach supported by a tool called Role-AdaptIA to detect and help to solve problematic situations in authentic computer supported collaborative learning (CSCL) scenarios. Role-AdaptIA is an adaptive interaction analysis tool based on a theoretical framework for the description of roles. The framework permits to define and characterize the roles to take into account in a given situation. Based on this information, Role-AdaptIA automatically detects role changes during the development of the collaborative experiences and warns the teacher about these changes. With this advice, the teacher is able to regulate the collaboration, providing support to the students to improve their interaction patterns. This chapter presents four examples of how Role-AdaptIA was used by teachers in order to regulate collaboration, detecting and solving undesired collaborative situations in several University courses where we have applied CSCL methods during the last years.

1 Introduction

Computer Supported Collaborative Learning (CSCL) is a research paradigm that underlines the key role that social interactions play in the process of learning [1].

In this process, particular forms of interactions among people are expected to occur, but there is no guarantee that these expected interactions are produced during the development of the collaborative experience [2]. There have been different proposals of techniques to increase the probability that some types of positive interaction occur, thus improving the collaborative process in general. These techniques are enacted at two moments: before the beginning of the collaborative experience and during the development of the experience [3].

Previous to the beginning of the collaborative experience it is necessary to design carefully the activity, taking into account the different aspects that might structure the learning experience, like the context, the group size, the group composition, the collaborative task or the definition and distribution of the participants' roles [2].

These pre-established conditions can change during the development of the activity due to the dynamism of learning in real contexts [4]. For example, the roles played by participants can vary, like a student emerging as a leader in a group, thus influencing the interactions in that group. Then, it might be necessary to warn the participants about these changes and to adapt the pre-established conditions to the new situation. This allows to regulate dynamically the activity and eventually, to improve the collaborative experience.

Regulation of collaborative activities can be support by Interaction Analysis (IA) methods and tools. These regulation processes aim to promote effective interaction using the results of the analysis to provide adequate feedback to the participants [5]. Then, the students, teachers or the system itself might recommend actions to help participants manage their interaction by reassigning roles, addressing conflicts and misunderstandings, or redistributing participants' tasks, given their levels of expertise [5].

Following these ideas we have proposed a role-based framework [6] and an adaptive IA tool called Role-AdaptIA (Role-Based Adaptive Tool for Interaction Analysis) [7] for supporting authentic collaborative experiences.

Role-AdaptIA applies social network analysis (SNA) [8] as a specific IA technique. SNA focuses on the study of social interactions. From a learning sciences point of view, these social interactions are related to the situated learning perspective [9], an approach to understand learning in authentic situations, which considers the social and cultural contexts where the experiences are produced, and that interprets learning as participation in the social world [10]. SNA, with its focus con social interactions, is appropriate for the study of these participatory aspects of learning [11]. Moreover, SNA techniques paired with recent developments in software for visualizations can help provide a clearer picture of what is happening in the online class. Thus, SNA appears as an appropriate technique for analysing learning in authentic CSCL scenarios.

Both the framework for the description of roles and Role-AdaptIA have been validated following an iterative design process, based on several case studies in real CSCL blended environments [12], [13]. In this context, a first experience evaluated the capacity to successfully adapt the results of Role-AdaptIA to the needs of the generic and pre-established roles of *student* and *teacher* [12] for different purposes (i.e., students' self-regulation and teachers' evaluation). A new experience [13] showed that the structured description of roles proposed in the framework provides appropriate information to describe and identify a set of roles (teacher-guide, teacher-observer, student-coordinator and student-isolated). In the present work, we describe the use of the framework and Role-AdaptIA to detect and help to solve potentially problematic situations in terms of collaboration, such as the existence of isolated students, or the malfunctioning of workgroups

Next section provides a brief explanation of the main characteristics of the framework and of Role-AdaptIA, as well as the main social network indexes taken into account in the collaborative experiences. Then, section 3 describes some examples of the collaboration improvements achieved using the proposed role-based approach. Finally the last section presents the main conclusions and an overview of our future work plans related to these items.

2 A Role-Based Interaction Analysis Approach to Support real CSCL Experiences

This section presents a brief summary of our proposal of a framework for the description of roles in CSCL and an adaptive IA tool for supporting authentic collaborative experiences.

The *framework for the description of roles* [6] enables the definition of the needs and characteristics of the different participants in a CSCL situation in terms of roles they can play in a specific context. The framework permits to define how to detect role changes during the collaborative activity, in terms of IA indicators. The description of a role in the framework is composed of four aspects: definition, context of application, IA information needs, and indicators for the dynamic detection of the roles previously defined.

The role's definition includes its name and the description of its functions. The role's name is divided into two parts: the *generic* name, which refers to the classical learning roles (e.g. teacher, student, or designer) and the *specific* name, which refers to the role's functioning in terms of social interactions (e.g., coordinator, leader, facilitator, isolated, semi-isolated, intermediary, observer or collaborator). The *context* specifies the type and size of the groups, the environment type, the educational level, the collaborative experience of the participants and the collaborative tools and tasks used in the scenario being analysed. The *IA information needs* aspect details the information required for a role in this context, which includes the information type, content, format and complexity, as well as the frequency and communication medium that will be used to send this information. Finally the *indicators for the dynamic detection of roles* aspect permits to define the indicators and values that will enable an IA tool to identify the emergence of a role during the activity.

As mentioned in the previous section, we used SNA in order to define these roles. More concretely, in the work presented in this chapter we employed the following ones: *actors' degree, closeness and betweenness centrality*; *network's density and centralization* and *sociograms*. They are briefly described below.

First, we considered three complementary indexes of actor's centrality: *degree, closenness and betweenness centrality*. *Degree centrality* ($C_D(i)$) measures the activity of an actor in the network, which gives an idea of his/her level of participation in it. Also, it is an index of the actor's prestige [14]; *closeness centrality* ($C_C(i)$) specifies the proximity of an actor to the rest of actors in the network. This index can be interpreted as a measurement of the influence of an actor in the overall network. Finally, *betweenness centrality* ($C_B(i)$) identifies the actor's position in terms of its ability to make connections to other actors in a network. This measurement represents the actor's capability to influence or control interactions between actors that are not directly linked.

We also introduced network indexes, such as *density* ($D(i)$), which is the ratio between the number of current links in the network and the total number of possible links. This index measures the intensity of activity in the network. Finally, *centralization* ($C(i)$) is another network index that measures the degree up to which the network activity (number of links) depends on a reduced number of actors.

In the case of relationships that consider the direction of the link, two degree, centralization and closeness indexes are defined. For example, for $C_D(i)$: *indegree* ($C_{Di}(i)$), or the number of links terminating at the node; and *outdegree* ($C_{Do}(i)$), or the number of links originating at the node.

Sociograms were also selected for the visualization of the detected roles. The sociograms represent the actors as nodes and the relationships among them as lines in the graph. Using specific node-positioning algorithms it is possible to identify roles in a very intuitive fashion. For example, a centred node in the graph points to a prominent actor, while a peripherical node denotes an isolated or semi-isolated actor.

As mentioned beforehand, the framework is supported by a tool, called *Role-AdaptIA* [7]. This tool checks automatically the different participants' roles during the activities, and detects role changes on the basis of the indicators and values that characterize each role, which are set previously by the teacher using the structure defined by the framework. When the tool detects a role change, the teacher receives a warning. With this information and other contextual data she might consider, the teacher can confirm (or not) this new role. In the first case, Role-AdaptIA adapts dynamically the information provided to the corresponding actor according to the needs specified for this new role. Currently, Role-AdaptIA draws on SAMSA [14], an IA tool that takes as an input interaction data coming from the logs provided by the CSCL tool. It builds social networks representing the interaction among the users and computes the mentioned SNA indexes.

As mentioned beforehand, the automatic detection of role changes carried out by Role-AdaptIA is motivated by the need to support teachers in regulating collaboration in real CSCL scenarios. Next section describes four examples that illustrate this approach.

3 Detecting and Solving Negative Situations during the Collaborative Experience

This section describes how the framework and Role-AdaptIA were used to identify negative situations during the development of CSCL experiences. First, we present briefly the main characteristics of the learning scenario where these experiences took place, and then we describe four examples of how we detected and solved different types of undesired collaboration situations.

3.1 Context of the Experience

The experiences reported in this section are part of a case study carried out since February 2005 in several courses of "ICT (Information and Communication Technologies) applied to Education" at the University of Valladolid. The setting is a blended one, where face-to-face activities are interleaved with in-site or distance technology supported activities. The participants collaborate mainly using Synergeia [15], a tool that provides a workspace for sharing documents among all participants in the course. Each course is divided into two main phases. First, a *theoretical phase*, during which students have to analyze different aspects of the subject matter and elaborate three reports collaboratively, and second, a *practical phase*, during which students have to create a Webquest, which could be eventually used in a real school. Students did not have previous experience in collaborative learning using computers, while teachers were familiarized to the use of interaction analysis tools for evaluation.

Previous to the beginning of the experiences, the teacher defined the set of roles to detect and the type of information that should be given to each of these roles. This information was then used by Role-AdaptIA to automatically detect changes in roles. Some of these changes are a symptom of a negative situation, and the message sent by the tool can be considered as an alarm, that warns teachers so that they can react on time to face that potentially problematic situation. Some of the roles defined were intrinsically negative, i.e., they represent undesired behavior with respect to collaboration. For example, the *student-isolated* role, a student that does not collaborate with the rest of participants. Other roles that can not be considered negative in principle, such as the *student-coordinator* could help to detect undesired situations when analysed in context, as it will be shown later in this section.

In the rest the section we describe a set of the collaboration problems detected using Role-AdaptIA. For this aim, we studied the role changes detected by the tool and we collected the warnings sent to the teacher in order to identify the negative collaboration situations. Then, we analysed the evolution of these situations after the teacher's intervention, which allowed us to assess the improvements achieved. For each one of the cases presented, we describe the role changes identified by Role-AdaptIA, the collaboration problems associated to these changes, the warnings sent to the teacher and the evolution of these problematic situations after the teacher's intervention.

3.2 Isolated Students

The first case we will present relates to the detection of students that do not participate in the collaborative activities. For this, we defined the *student-isolated* role, which is characterized by an absolute lack of interactions. In terms of the SNA indexes defined for its detection, this role is characterized by a value of zero in the $C_D(i)$ indexes (indegree $C_{Di}(i)$, and outdegree $C_{Do}(i)$). Besides, the students with this role appear disconnected to the rest in the sociogram, i.e., they do not have links with other actors in the network.

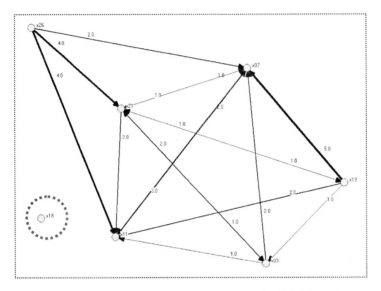

Fig. 1. Sociogram representing an isolated student x18 within his workgroup

An analysis of the activity within the working groups defined in this course (intra-group analysis) performed by Role-AdaptIA two weeks after the beginning of the semester identified a case of student-isolated corresponding to actor *x18* with $C_{Di}(x18)$ and $C_{Do}(x18)$ equal to zero. Figure 1 shows the sociogram associated to this intra-group analysis, where it is easy to see that node *x18* is not linked with the members of his/her group.

Additionally, the analysis of the global activity of the students with respect to the rest of participants in this course during these two weeks showed the same results for this student, i.e., $C_{Di}(x18)=0$, and $C_{Do}(x18)=0$. The sociogram associated to this analysis (see Figure 2) showed clearly this fact, displaying *x18* with no connections to the rest of the actors in the network.

After the identification of this role, Role-AdaptIA sent a warning to the teacher, as displayed in Figure 3. It is possible to see that the warning included information about the detection of a student-isolated role associated for student *x18* into his/her group and with respect to the rest of the classroom. Besides, the warning message included the values of *x18*'s SNA indexes as well as the two associated sociograms.

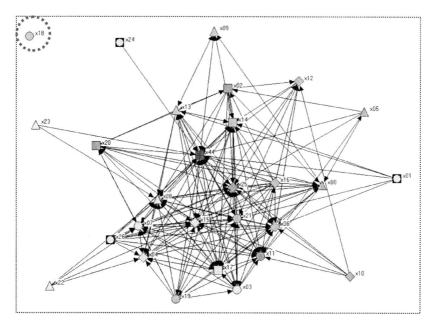

Fig. 2. Sociogram representing the individual activity of all participants in the course after two weeks from the beginning of the experience. Different node shapes are used for each working group.

After receiving this information, the teacher interviewed $x18$'s group members, in order to gain insight into the conditions leading to these indexes. It was found that $x18$ was not attending lectures, as it was later confirmed by the observations performed by an external observer. Besides, $x18$ did not interact off-line with his partners through Synergeia.

A new analysis performed by Role-AdaptIA showed two weeks later that the student $x18$ had changed his behaviour. At that moment, his indexes were $C_{Di}(x18)=4$, and $C_{Do}(x18)=16$. Although this student could not assist to lectures, he had started to work remotely, using Synergia in order to perform the collaborative tasks assigned by the teacher.

With this case we have illustrated a first example of the detection of an undesired student's role. The warning sent to the teacher by Role-AdaptIA and the intervention of the teacher allowed to modify this negative situation and to improve the collaborative activity. Although it might appear that the identification of such a student-isolated role is rather straightforward, the fact is that in this scenario it was difficult for the teacher to identify which students were not interacting at all (either face to face or remotely through Synergeia). Thus, the warning sent by Role-AdaptIA enabled the teacher's intervention and the improvement in the behaviour of the student.

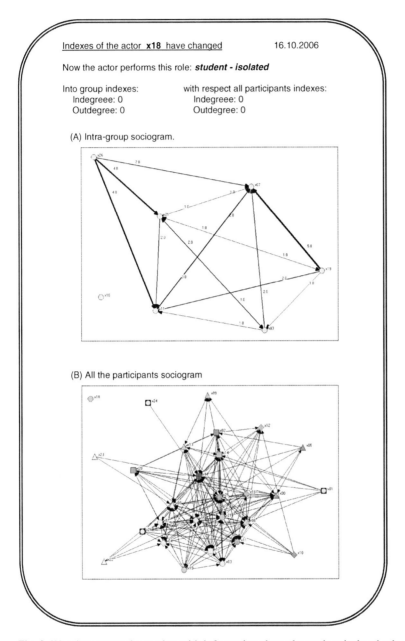

Fig. 3. Warning sent to the teacher with information about the student-isolated role detected

3.3 Intra-group Conflicts

In this case, Role-AdaptIA detected a case of intra-group isolated student. After performing an intra-group analysis, it was found that student *x15* did not interact

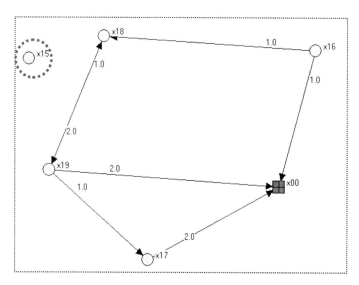

Fig. 4. Sociogram representing "x15" as an isolated student within her work group

with the members of her group, i.e., her indexes were $C_{Di}(x15)=0$, and $C_{Do}(x15)=0$. Figure 4 shows the sociogram associated to this intra-group analysis, where it is easy to see that *x15* appears isolated from the rest of the group, including the teacher, represented by a square.

On the contrary, the general analysis of the activity of each student with respect to the rest of the classroom did not identify *x15* as an isolated student, with $C_{Di}(x15)=5$, and $C_{Do}(x15)=3$. The sociogram associated to this general analysis showed that *x15* was linked to four members of other groups. Figure 5 shows this sociogram.

The teacher received a warning about the detection of a student-isolated role into Group 5, together with the sociogram displayed in Figure 4, and her values $C_{Di}(x15)=0$, and $C_{Do}(x15)=0$. Besides, Role-AdaptIA sent the sociogram associated to the analysis of the individual activity (Figure 5).

A conflictive situation had been discovered, but not the reason. In contrast to the case discussed in the previous section, this student was attending lectures and participating in the activities, but, according to the information provided by Role-AdaptIA, she was not working with her group mates. The teacher interviewed the components of the group, and it was found that there was a personal conflict between two members of this group. A similar conflictive situation was detected into another group, where the student *x33* did not collaborate with the members of his group. He only interacted with the student *x18*, a member of a third group. We can see this link between *x33* and *x18* in Figure 5, on the lower part of sociogram. Role-AdaptIA detected also this fact and warned the teacher about it. In order to solve the conflictive situation the teacher decided to restructure the composition of the affected groups, with a rearrange where *x33* was assigned to *x15*'s former group and vice versa.

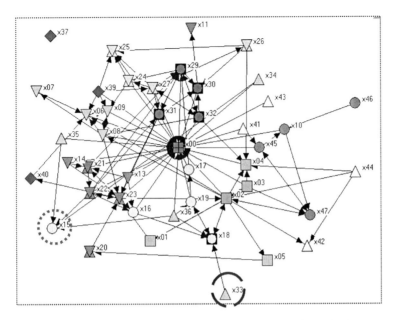

Fig. 5. Sociogram representing individual activity of all participants in the course

Next analysis performed by Role-AdaptIA showed that both groups did not contain isolated students anymore. Thus, the intervention of the teacher with the support of Role-AdaptIA had been successful in reverting this negative situation.

This conflictive situation was difficult to detect on-line by the teacher, because the affected students did participate in the classroom activities, although not with their group mates, due to the mentioned personal conflicts. The high number of students in each classroom, as it is usual at university settings, makes these internal conflicts difficult to detect. In this case, Role-AdaptIA helped the teacher to detect this problem, which enabled the intervention that eventually improved the internal dynamics of the involved groups.

3.4 Group Malfunctioning

In the previous cases, the detection of the negative situation (isolated students and intra-group conflicts) was based on the detection of a negative role. Additionally, Role-AdaptIA also helps to identify undesired situations hidden behind the detection of non-negative roles, as it happened in the example described in this section, where the detection of a *student-coordinator* role allowed identifying a group that did not collaborate properly.

A *student-coordinator* is defined as a student that organizes the activity into the group. He/she regularly initiates the interaction with the rest of his/her group members. Besides this, he/she maintains relationships with other groups. Table 1 shows the SNA indicators and values defined in this case by the teachers for the on-line detection of the student-coordinator role, according to the structure given by the framework.

The indicators are the ones defined in section 2. The relevance rank column states the weight of the indicator for detecting this role, specified as a priority rank (e.g., first, second). Role-AdaptIA confirms the detection of an emergent role only if all the indicators associated to a first level of relevance rank verify the specified values. This is a mandatory condition in order to validate the detection of a role-change

During the development of the collaborative activity, Role-AdaptIA detected an emergent *student-coordinator* role into a group (Group 8). The values of the SNA indexes of this student, *es18*, verified all the specifications established for this role (see Table 2) with $C_{Di}=100$, $C_{Do}=100$, $C_B=100$.

Moreover, the associated sociogram (see Figure 6) shows how the student *es18* was the most centred node. These conditions indicated that *es18* was a student-coordinator of this group.

Table 1. Specification of the indicators and their values for the student-coordinator role

Role: student – coordinator			
Indicators	*Values*	*Interpretation*	*Relevance*
Outdegree $C_{Do}(i)$	> 75%	A very high value indicates that the student controls the work of the group members	First level
Outcloseness $C_{Co}(i)$	> 75%	A very high value indicates the high proximity of the student to the rest of the group members	First level
Betweenness $C_B(i)$	> 75%	A very high value indicates good connections to the rest of the group members	First level
Actor position in sociograms	Near the center	A centered node in the graph indicates a prominent actor for the rest of participants	Second level (for visual validation)

Additionally, it is easy to see in Figure 6 that the other group members did not present interactions among them. They only had interactions with *es18*. These students had the degree centrality indexes very low or zero. Besides, their degree sessions and activity values were very low. They were identified as *students-semi-isolated* or *students-non-participatory*.

The teacher received warnings about these role detections together with the values of the related SNA indicators for this group and the associated sociogram. This information suggested abnormal collaboration.

In order to understand better this group's behaviour, the teacher interviewed the group members and reviewed the detailed information elaborated by Role-AdaptIA about the interactions into this group. The conclusion was that the students of Group 8 did not work collaboratively. They cooperated in order to

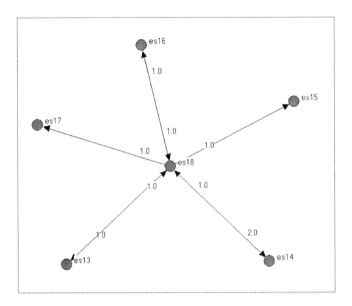

Fig. 6. Sociogram representing the interactions into Group 8. The centered *node*, *"es18"*, was the student-coordinator role detected

elaborate the report associated to the collaborative task dividing it into five parts. Each student elaborated one of them and later, *es18*, put together all the fragments, creating a final report. For this aim, *es18* read documents coming from the rest of the members (each part of the report), and some of these members read the final report created by *es18*. Even the students *es15* and *es17* did not read the final report "elaborated" by their group.

In order to improve collaboration in this group, the teacher dialogued with the group members in the classroom. The teacher explained the collaboration concept again and clarified why the work performed by Group 8 was not real collaboration.

Next analysis performed by the IA tool allowed to verify how the intervention of the teacher helped to improve the intra-group collaboration. During the next task, i.e., the elaboration of the second report in the practical phase (see section 3.1) the analysis showed the evolution of the interactions into Group 8. Figure 7 shows the sociogram associated to this task. We can see that all the members are linked with several partners (three, four or five), that is, they have interactions among them.

The undesired situation presented in this section probably would have gone unnoticed on-line by the teacher. The warning sent by Role-AdaptIA and the information attached was fundamental in order to detect a group that did not collaborate properly, and the teacher's intervention achieved to improve the collaborative activity into the group.

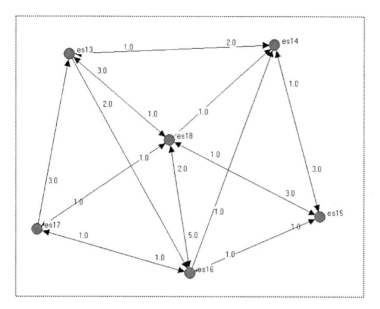

Fig. 7. Sociogram representing the evolution of the interactions into Group 8

3.5 Teacher Dependent Students

The three negative situations shown above were based on the detection of a student-role change identified by Role-AdaptIA. In this section, we present an undesired collaborative situation based on the analysis of the teacher role changes. In this case, the initial detection of a *teacher-guide* role and the evolution towards a *teacher-collaborator* role helped to identify students that were totally dependent on the teacher's interventions and suggestions. These students did not improve their autonomy during the collaborative experience.

The analysis of the teacher's activity during the development of the tasks detected a role change. Initially, during the elaboration of the first report by the students, i.e., in the theoretical phase, a *teacher-guide* role was detected. According to the SNA indexes, she conducted the activity and her contributions were the most relevant for the students. Her indegree and incloseness were the highest of participants (29 and 10.57 respectively) and she was the most centred node in the sociogram. This was expected by the teacher, because the students had not previous experience with collaborative work and tools.

Just as it was also expected, the teacher's relevance decreased during the next weeks. Thus, after the first part of the practical phase Role-AdaptIA detected that the teacher had lost her role of a *guide* and she had become a *collaborator*. More than 50% of students presented higher values in their SNA indexes (with C_{Di} values ranging from 132.00 to 21.00 and C_{Ci} values from 10.39 to 8.82, while the teacher's values of these indexes were $C_{Di}(teacher)=14.00$ and $C_{Ci}(teacher)=8.77$, respectively), and her position in the sociogram was not centred anymore. These indicators reflected that the teacher was monitoring the activity but only participating at specific moments, for example for reading the reports elaborated by students in the theoretical phase.

The evolution detected on the teacher's role indicated that the students needed more the teacher's interventions at the beginning of the activity, because they did not know the subject and they had no previous experience with collaborative tools and tasks. Later, the students had reached certain autonomy and they could work more independently. At that moment was when the *teacher-collaborator* role emerged.

However, the intra-groups analysis performed by Role-AdaptIA indicated that the teacher behaved still as a leader for some groups. Then, the teacher received two types of warnings by Role-AdaptIA. On the one hand, Role-AdaptIA sent information about the groups where the teacher evolutioned to as a teacher-collaborator role. For example, Figure 8 shows the sociogram sent to the teacher associated to Group 2's activity during the first part of the practical phase. We can observe that the teacher, represented by a square node was not a centred node anymore. Moreover, the relationships among the members of the group do not vary if the teacher's node is excluded of the sociogram. The members of this group did not depend on the teacher on their workgroup.

On the other hand, Role-AdaptIA sent to the teacher information about the groups where she maintained the *teacher-guide* role. For example, Figure 9 shows the sociogram associated to Group 3's activity during the first part of the practical phase. We can observe that the teacher is the more centred node in the sociogram.

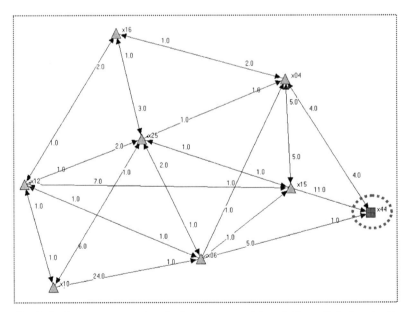

Fig. 8. Sociogram representing the intra-group activity of Group 2 during the first part of practical phase, including the teacher, represented by a squared node

The majority of Group 3's members were linked to the teacher, and they had very few links among them.

Figure 10 shows the activity of this group during the same period excluding the teacher. It shows a disconnected group, where each student is linked only to one or two other group members. This meant that the students were totally dependent of teacher's contributions in order to achieve a good collaboration, but they did not work collaboratively among them.

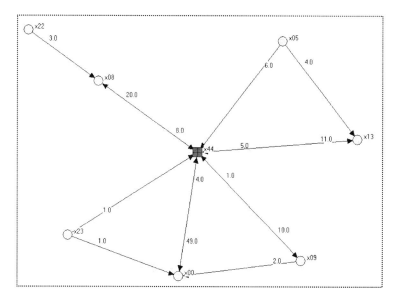

Fig. 9. Sociogram representing the intra-group activity of Group 3 during the first *part of practical phase, including the teacher, represented by a squared node.*

The teacher intervened discussing with these students about the lack of collaborative activity found in their group. She encouraged the students to increase this activity. A later analysis performed by Role-AdaptIA in this course allowed to verify that the teacher had lost her role of a *guide* in this group at the end of the practical phase, i.e., she had succeed in promoting the group's internal collaboration.

The undesired situation presented in this section was detected by analysing the expected evolution of the teacher's role from *guide* to *collaborator*. This evolution implied that the experience acquired by the students during the first phase of the course would allow them to achieve a certain autonomy with respect to the teacher during the following phases. The analysis of the groups where this evolution had not happen allowed to detect which students were still too dependant on the teacher's contributions. Then, she could intervene in order to face this lack of autonomy.

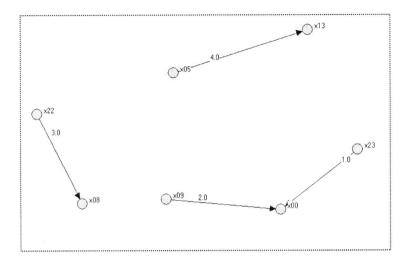

Fig. 10. Sociogram representing the intra-group activity of Group 3 (excluding *the teacher*), *during the first part of practical phase*

4 Conclusions and Further Work

IA methods and tools can help to regulate CSCL activities providing adequate feedback to the participants. IA helps to compare the actual collaboration state with respect to the planned before the beginning of the activity. Then, the teachers can intervene for recommending actions to the participants in order to improve the collaborative process.

This work has presented four negative situations identified in a real CSCL scenario using an approach based on the detection of role changes during the development of collaborative learning activities. Our approach uses a theoretical framework for the description of roles in CSCL scenarios, supported by an adaptive IA tool called Role-AdaptIA. With the structured description of roles given by the framework, Role-AdaptIA is able to adapt its output to the roles emerging during a collaborative learning activity, and this way, it helps to regulate collaboration in the classroom.

In the cases described in this work, the process for the detection of negative situations began with the identification by Role-AdaptIA of a role change during the collaborative activity. Then, Role-AdaptIA sent a warning to the teacher advising about the new role detected. Besides, the teacher received SNA information (numerical and graphical) associated to this role change. This information helped the teacher, either directly or indirectly, to discover negative collaboration situations. On the one hand, some negative roles, such as the student-isolated, allowed to identify undesired collaboration situations. On the other hand, some roles not considered as negative in principle, covered conflictive or negative situations, such as the student-coordinator role case. The detection of this role helped to detect a workgroup that was not collaborating properly. Finally, Role-AdaptIA also supported the identification of other negative situations by the analysis of the teacher's role changes. In all these cases, the warnings received by teachers and their later interventions achieved to solve these negative situations, and eventually, to improve the collaborative activities. It is important to note that without the support provided by Role-AdaptIA most of these situation would have went unnoticed by the teacher.

Upcoming work plans point to incrementing the number of roles to identify by Role-AdaptIA as well as to extend the tool functionality to allow the teacher new and finer types of analysis. Overall, these aspects are meant to contribute to improve the regulation of computer-supported collaborative learning activities in real practice.

Acknowledgments. This work has been partially funded by the Spanish Ministry of Education and Science (TSI-2005-08225-C07-04) and the Autonomous Government of Castilla y León, Spain (projects VA107A08 and VA106A08). The authors would like to thank the rest of GSIC/EMIC group at the University of Valladolid for their support and ideas.

References

1. Koschmann, T.: Paradigms shift and instructional technology. In: Koschmann, T. (ed.) CSCL: Theory and practice of an emerging paradigm, pp. 1–23. Lawrence Erlbaum, Mahwah (1996)
2. Dillenbourg, P.: Introduction; What do you mean by "Collaborative Learning?". In: Dillenbourg, P. (ed.) Collaborative Learning. Cognitive and Computational Approaches, pp. 1–19. Elsevier Science Ltd., Oxford (1999)
3. Jermann, P., Soller, A., Lesgold, A.: Computer software support for CSCL. In: Dillenbourg, P., Strijbos, J.W., Kirschner, P.A., Martens, R.L. (eds.) Computer-supported collaborative learning: What we know about CSCL.. and implementing it in higher education, vol. 3, pp. 141–166. Kluwer Academic Publishers, Boston (2004)
4. Edwards, W.K.: Policies and Roles in Collaborative Applications. In: Proceedings of the 1996 ACM Conference on CSCW, pp. 11–20. ACM Press, New York (1996)
5. Soller, A., Martínez, A., Jermann, P., Muehlenbrock, M.: From Mirroring to Guiding: A Review of the State of the Art Technology for Supporting Collaborative Learning. International Journal on Artificial Intelligence in Education (15), 261–290 (2005)
6. Marcos-Garcia, J., Martínez-Monés, A., Dimitriadis, Y., Anguita-Martinez, R.: A role-based approach for the support of collaborative learning activities. e-Services Journal 6(1) (2007)
7. Marcos-García, J.A., Martínez-Monés, A., Dimitriadis, Y., Rodríguez-Triana, M.J.: Role-AdaptIA: A role-based adaptive tool for interaction analysis. In: ICLS workshop, Utrecht, The Netherlands, June, 23-24 (2008)
8. Wasserman, S., Faust, K.: Social Network Analysis: Methods and Applications. Cambridge University Press, Cambridge (1994)
9. Wenger, E.: "Communities of practice. In: Learning, meaning and identity, Cambridge University Press, Cambridge
10. Sfard, A.: On two metaphors for learning and the dangers of choosing just one. Educational Researcher 27, 4–13 (1998)

11. Scott, J.: Social Network Analysis: A handbook, 2nd edn. Sage Publications, Thousand Oaks (2000)
12. Marcos, J.A., Martínez, A., Dimitriadis, Y., Anguita, R.: Adapting interaction analysis to support evaluation an regulation: a case study. In: Proceedings of the 6th International Conference on Advanced Learning Technologies, ICALT 2006, Kerkrade, The Netherlands, July 2006, pp. 125–129 (2006)
13. Marcos, J.A., Martínez, A., Dimitriadis, Y.A., Anguita, R.: Interaction analysis for the detection and support of participatory roles in CSCL. In: Dimitriadis, Y.A., Zigurs, I., Gómez-Sánchez, E. (eds.) CRIWG 2006. LNCS, vol. 4154, pp. 155–162. Springer, Heidelberg (2006)
14. Martínez, A., Dimitriadis, Y., Tardajos, J., Velloso, O., Villacorta, M.: Integration of SNA in a mixed evaluation approach for the study of participatory aspects of collaboration'. In: European Conference on Computer Supported Collaborative Work (ECSCW 2003), workshop on Social Networks, Helsinki, Finland (2003)
15. ITCOLE Research Project. About Synergeia (retrieved November 2008), http://bscl.gmd.de

DyMRA: A Decentralized Resource Allocation Framework for Collaborative Learning Environments

Xavier Vilajosana, Daniel Lázaro, Joan Manuel Marquès, and Angel A. Juan

Universitat Oberta de Catalunya, Rambla Poblenou 156 08018 Barcelona
e-mail: {xvilajosana,dlazaroi,jmarquesp,ajuanp}@uoc.edu

Abstract. Collaborative e-learning virtual communities use virtual learning environments provided by the university or tools that are available in Internet. Although this model has proven to work, it has important limitations. A way to deal with them is by defining virtual organizations (VO) that gather the resources and interests of their members in a way that they can lend resources to or borrow them from other VO. This paper presents DyMRA, a decentralized resource allocation system based on markets that allows inter-VO resource allocation. DyMRA is specially designed for collaborative peer-to-peer environments, where the autonomy of participants and its decentralized nature requires the capacity to dynamically reallocate resources and services that manage the overall system. DyMRA is built on top of La-COLLA, a peer-to-peer middleware that allows a group of users to share resources in a collaborative manner. We present the design, architecture and validation of our proposal.

1 Introduction

Nowadays nobody doubts that E-Learning is an effective and useful way to learn, as demonstrates the success of many virtual universities – the Open University of Catalonia (http://www.uoc.edu), with its almost 40.000 students, is an example of it – or the fact that, in traditional universities, many subjects are taught totally or partially in a virtual manner. Collaborative e-learning has also deserved a lot of attention and many solutions have been proposed. In either case, the resulting virtual communities use virtual learning environments provided by the university which helps to preserve the community notion – or using tools that are available in Internet – either for free or paying. Although this model has proven to work, it has important limitations, like the following three: learning institutions has to estimate the amount of resources needed to deal with peak situations, with the resulting overdimension of the system; different systems (e.g. belonging to different faculties) do not share resources, which may result in having in the same university some systems overloaded at some periods of the year while other systems, at this period, are underutilized; and that ad hoc collaborative groups are only partially supported, because the tools and resources available to them are restricted to university capacity and polices.

A way to deal with those limitations is by defining virtual organizations (VO) that gather the resources and interests of their members in a way that they can lend resources to or borrow them from other VO. Examples of VO can be a faculty department or a group doing a collaborative activity. Whilst the resources within a VO may be sometimes insufficient to satisfy QoS requirements under unexpected load surges, high level of dynamicity of its members or unpredictable failures, computers that belong to another VO may have surplus bandwidth, storage and computations resources. This opens challenging opportunities to promote inter-VO resource allocation. In other words, a VO could aggregate their surplus resources and offer them to other VOs.

In this paper we address Inter-VO allocation of resources by means of decentralized markets that promote the creation of local ad hoc trading sites that can be accessed by any VO. The reason to use markets for the resource allocation is because they have proven their ability to allocate resources efficiently [1] and, more importantly, they provide mechanisms through which the need may be correctly elicited and quantified; and indeed, they promote incentives to resource owners to provide or trade their resources.

We developed DyMRA, a decentralized resource allocation system based on markets that allows inter-VO resource allocation. DyMRA is specially designed for dynamic and peer-to-peer environments, where the autonomy of participants to disconnect resources at any time and its decentralized nature requires the capacity to dynamically reallocate resources and services that manage the overall system. Besides, DyMRA can be seen as the middleware to promote contributory environments. Contributory environmnets are those where users provide their own resources to be used collectively either for free or expecting some revenue. We claim that collaborative applications can be supported by contributory environments and indeed contributory environments have the characteristics that make them adequate for collaborative applications.

In DyMRA markets are created at will and run as services within the VO. The choice of a decentralized markets approach in the form of many local ad hoc markets is motivated by the need to deal with dynamic communities and scalability issues that would be limited by a centralized approach.

In the scope of our research a resource specification and bidding language have been developed and used by DyMRA components. For the purpose of this paper, we consider resources as processing time, storage, and applications that provide a stateless service, like an efficient codec or a parallelizing compiler. We deal with heterogeneity by using standard interfaces, implemented as Web Services, to access the resources. These belong to a VO and we assume that they can be disconnected or fail at any time. This dynamic behavior introduces a complexity that, added to the decentralized behavior of markets, forced us to design a system that allows us to decouple services from physical resources. Furthermore, we had to pay special attention to availability guarantees.

Another important aspect in a system that allows allocation of external resources is that of defining a medium of exchange, namely a virtual currency. A virtual currency facilitates the transfer of services or resources. Furthermore, the virtual currency rather than storing value act as a token that simplifies preference elicitation and provides incentives to share resources while facilitating trading without incurring in real payment. In some cases, may be interesting that virtual currency could be translated into real money. In some environments, though, this might not be desirable. For example, in the case of VOs formed inside a university, resources should be shared without needing to exchange real money. Hence, we use virtual currency as a means to quantify the resources a VO can allocate, leaving the option to translate this into real money for the environments where it's appropriate. In DyMRA, currency is used as a regulatory mechanism.

DyMRA is built on top of LaCOLLA[1] [2, 3]. LaCOLLA is a peer-to-peer middleware that allows a group of users scattered across the Internet to share resources in a cooperative manner and that allows the deployment of stateless services using the resources provided by the members of the VO. LaCOLLA guarantees that services deployed are always available (if enough resources are provided). Therefore, DyMRA components are deployed as services in LaCOLLA middleware.

The rest of the chapter is organized as follows: Section 2 presents a scenario to better illustrate our proposal and clarify the issues to be addressed in this paper. Section 3 presents the requirements of our system. Section 4 relates DyMRA to other works. Section 5 describes the architecture and the behavior of main processes. Section 6 present the approach taken for the market repository, section 7 and 8 introduce the agreement and payment methodologies used in our prototype. Section 9 discusses the type of markets that can be used in DyMRA while section 10 economically evaluates the presented approach. Finally Section 11 presents the evaluation of the mechanisms used in DyMRA while section 12 concludes with an outlook to future work.

2 Scenario

A community formed by cookery book readers creates a VO (VO-1) whose objective is to share knowledge and teach each others some "recipes from the days of sail". The technical objective of the VO-1 is to provide the storage and processing capacity to store the data generated by VO-1 members applications. One application (A) digitizes video and transforms it to a computer-readable format. A second (B) adds subtitles into each video file, while the last and popular application (C) is a real-time player that plays the videos and diffuses multimedia content to other VO-1 members. The VO-1 has a large number of subscribed members most of who contribute sporadically their resources. All the three applications require processing time and the first requires storage as well. While a few members contribute regularly their computational resources to the VO-1, the majority provide them sporadically.

---

[1] Available at http://dpcs.uoc.edu/lacolla/.

They instead pay a subscription fee to obtain this service. In this scenario, the focus of this chapter is the allocation of resources to the applications. We assume that all resources and applications of the VO-1 are managed and the management logic takes appropriate decisions to ensure that preset goals are met. If this decision triggers resource allocation, then a DyMRA component adopting the role of a buyer agent negotiates at the market place to acquire resources.

At a time any of the applications A, B, C may have load surges and require resources to match the required quality of service. Application (A) requires both storage and processing time since the video must be digitized and stored. The execution of the digitizer is planned and scheduled by the VO administrator and hence resources are leased in advance. Application (B) may have unplanned load surges due to remote requests by VO members to subtitle video files. Allocation of resources for this application is triggered by the load monitoring logic, but members may be requested to wait. The buyer agent within DyMRA looks for a market that trades in processor cycles that is available within the time range as required by the application. If the Market Directory does not return matching running (and advertised) market instances, the buyer agent instantiates a new auction market. Application (C) is stringent in its resource requirements and cannot wait for allocation. When it requires more resources the buyer agent will query the Market Directory for a continuously clearing double auction that trades in processing time for immediate usage.

Other VOs, for example VO-B and VO-C are formed by a numerous set of university members and have surplus resources. In order to obtain a benefit from their unused resources they offer some of their computational capacity and storage resources to other VOs through DyMRA markets.

DyMRA addresses this kind of scenarios by providing services to automatically allocate external resources into a VO. The main contributions of DyMRA are twofold; first, the components of DyMRA are deployed as services inside a VO and can, hence, be reallocated when its current location fails or disconnects, keeping the functionality available. Second, DyMRA proposes to distribute markets amongst virtual organizations that place our approach in a design space between a decentralized and a centralized architecture, which we believe responds better to the targeted environment.

3 Requirements

- **Interoperability:** VO services may be exposed as standard interfaces that enable interoperation between VOs.
- **Group Self-sufficiency:** The execution of services and the deployment management should be performed using only the resources contributed to the VO by its members.
- **Decentralization and Self-organization:** In case of connections, disconnections and failures, the system should keep functioning (it shouldn't have a single point of failure) and should reorganize without requiring any external intervention,

getting to a consistent state as soon as the available resources and VO stability allow it.
- **Individual Autonomy:** The VO's members should be free to decide which actions to carry out, what resources and services to provide, and when to connect or disconnect.
- **Market Availability:** Market services should always be available (if needed) as long as there are enough resources to execute them in the VO.
- **Location Transparency:** Market services don't have to worry about other market service's location. The system resolves them transparently, and services access each other using a location-independent identifier.

4 Related Work

Economic based resource allocation within the context of Virtual Organizations, Grid Computing and large scale peer-to-peer systems has been extensively studied [4, 5, 6]. However, as far as we know, the issue of addressing inter-VO resource allocation is an emergent field of study. Recently in [7] a novel architecture for inter-VO resource allocation in Grids is presented. Their proposal is suited between a centralized and a decentralized approach and proposes a configurable mediator process (executing within the VO) that allocates resources from external providers. Our approach goes one step beyond and we provide, on one hand, transparent reallocation of market services in a dynamic environment, and, on the other hand, we rely on the members of collaborating communities to mutually provide resources.

Another feature that we addressed is that of dynamic deployment of services. We acknowledge some systems that perform this task in a decentralized and self-organized way, as is our target. Snap [8] nodes form a Distributed Hash Table (DHT) which stores the code and data of the services. Replicas of a service are created on demand and stopped when demand decreases. Another system called Chameleon [9] deploys services in a cluster of nodes communicated through a DHT, while trying to maximize its "utility" (calculated from a value assigned to each service and its performance in a given node).

5 Architecture

The architecture of DyMRA consists of a series of components which are in charge of the trading process. These components are:

- **Prospector:** when external resources are needed, it is in charge of finding a suitable market and obtaining the desired resources.
- **Seller:** it is in charge of offering the aggregated surplus resources of the VO in a suitable market.
- **Pool Service:** it controls the access of the VO members to the external resources acquired by the VO, acting as a mediator.

- **Sale Handler:** it controls the external access to a set of resources sold to another VO, acting as a mediator.
- **Accounting Service:** it monitors the resources available in a VO. Following a policy determined by the VO, it starts the acquisition of external resources or the cession of own resources to other VOs when convenient.
- **Market:** it mediates the trading of resources between VOs.
- **Market Directory:** Contains an index of existing markets and their locations.
- **Market Repository:** Contains descriptions and specifications of market mechanisms and configurations, ready to be deployed.

The system is built upon a middleware called LaCOLLA [2] which allows a small group of computers connected through internet to participate in collaborative activities and sharing their resources (i.e. provides virtualization of resources), while tolerating high levels of dynamism. This middleware also allows the deployment of services within a VO (or group) [3]. When a service is deployed, the system guarantees that it will always be available, placing it in a suitable location chosen among the resources of the VO, and reinstantiating it in case of failure.

The components of DyMRA are deployed as services inside a VO (except the Market Directory), and can, hence, be reallocated when its current location fails or disconnects, keeping the functionality available. The communication between VOs

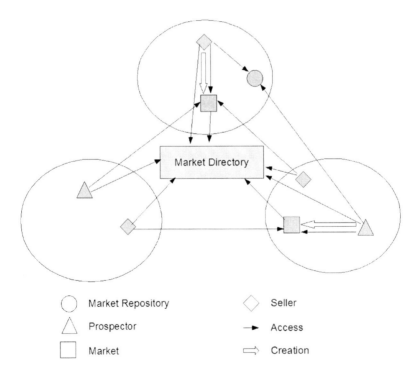

Fig. 1. Three groups trading resources through DyMRA

DyMRA: A Decentralized Resource Allocation Framework

is done through markets, which are also services, existing in a specific VO. To access a market, it must be discovered through the Market Directory (MD). Markets contact the MD to publish their location and characteristics, and the MD keeps them as a soft state. In case a market ceases to exist, the MD will delete the information about it after its time-to-live (TTL) expires. If a market is reallocated, it will inform the MD of its new location.

The MD is not part of a VO, but an external service which is known and can be accessed by all groups. Its implementation is out of the scope of this discussion, but there are many possibilities. It could be a centralized index, but it could also be implemented in a decentralized way if each VO deployed a "MD node" service, and each one of these services act as a node of a DHT, thus distributing the information stored among the VOs. Anyway, this doesn't affect the design of our architecture.

To help understand the functionality of each of the components presented and the overall behavior of the system, we will explain in detail how the trade of resources is done at the buyer VO and at the seller VO (shown at fig. 2), and how the posterior access to the traded resources is managed (fig. 3).

5.1 Trading Process

Buying resources

1a. The Accounting service detects that the resources of a certain type (e.g. storage) available in the VO are below a certain threshold defined by the VO policy. According to a given policy, it determines the resources needed and other factors such as the price that should be paid for them. With this information, it contacts the Prospector and asks it to acquire such resources.
2a. The Prospector looks for a suitable market in the Market Directory.
3a. The Market Directory sends the Prospector a list of markets which suit the specified needs.
4a. The Prospector chooses one of the markets of the list. In case that there is no suitable market, it proceeds to the creation of a new one. Once it has the adequate market located, the Prospector sends its bid. A generic bid describes the type of resource to bid for, the price per unit offered and the number of units required amongst others.

Selling resources

1b. The Accounting service detects that the resources of a certain type (e.g. storage) available in the VO are above a certain threshold defined by the VO policy. According to a given policy, it determines that these resources can be leased to another group, and fixes the price that should be paid for them. With this information, it contacts the Seller and asks it to sell the surplus resources.
2b. The Seller looks for a suitable market in the Market Directory.
3b. The Market Directory sends the Seller a list of markets which suit the specified needs.

4b. The Seller chooses one of the markets of the list. In case that there is no suitable market, it proceeds to the creation of a new one. Once it has the adequate market located, the Seller sends its offer.

Agreement

5. The market makes an agreement between the buyer and the seller. A scheduled double auction is used to match winning bids and offers. The winners are selected by calculating the price where supply balances demand and matching the highest buy bids above the price with the lowest sell offers below the price. After this, it notifies the sale to both the Prospector and the Seller.
6. The Seller starts a Sale Handler, which is deployed in its VO. This Sale Handler keeps the information about the leasing conditions, and mediates the use of the resources according to these conditions.
7. The Prospector informs the Pool service of its group about the resources bought and the agreement conditions, as well as the location of the Seller of the resources.

When a Prospector or a Seller finds that there is no market available that suits its needs, it proceeds to the creation of a new one. As stated before, the market is implemented as a service. Hence, the component (Prospector or Seller) creates a new service in its VO, which is a market with the desired characteristics. This market registers itself in the Market Directory, and therefore can be accessed by buyers or sellers from outside the VO.

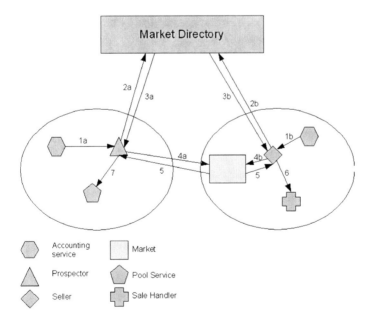

Fig. 2. Interaction among components in the trading process

5.2 Accessing the Resources

1. Whenever a client needs to use a resource, the system checks the VO policies to determine whether it must depend only on local resources or should use external resources. In the latter case, the client contacts the Accounting service.
2. The Accounting service checks the resources currently available to the VO. Following the VO's policy, it determines what resources the client must use, whether these are internal or external. In the former case, it tells the client which resource to use. Otherwise, it tells him to contact the Pool service.
3. The client contacts the Pool service, as if it was a local resource.
4. The Pool service chooses which of the external resources available to the VO should be used, and contacts its corresponding Seller. It sends the id of the sale it wants to use.
5. The Seller tells the Pool service the location of the Sale Handler that manages the specific agreement.
6. The Pool service contacts the Sale Handler, according to the conditions of the agreement, which may include, for example, symmetric key cryptography. It basically resends the request of the client.
7. The Sale Handler checks that the request of the Pool service does not violate the conditions of the agreement. After this, it uses the resources of the VO to fulfill the request of the Pool service.

As stated before, the services can change their locations due to failures or disconnections. This is not a problem inside a VO, as the system guarantees that clients, as well as other services, can contact any active service. To access external resources, though, the Pool must contact the Seller of another VO, whose location might have changed from the moment when the agreement was made. This can be solved in more than one way. A solution would be to use the Market Directory to store also the location of the Seller of each VO. This information would be maintained in a

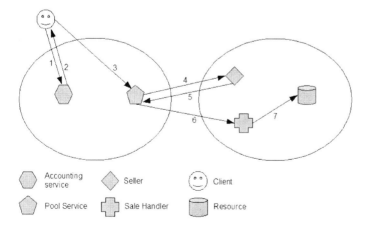

Fig. 3. Interaction among components in the access process

soft state, just like the one about markets, with the Sellers explicitly publishing their locations in the Directory. The Pool could then contact the MD to get the current location of the Seller, in case it cannot reach it in its previous location. This would solve the problem, but implies relying in an external entity (even though, as seen before, the MD can be implemented cooperatively by the VOs). A solution that only depends on the two VOs that need to communicate would be that both the Pool and the Seller keep the location of those Sellers and Pools, respectively, they have a deal with. In case one of these services is reallocated, it would notify all its "business partners" about its new location. Although it would be less probable, contact can still be lost if both Pool and Seller are reallocated at the same time. To further diminish this probability, these services could be replicated inside the VO. In the worst case, if all the replicas of both services fail together and the Pool of one VO can't contact the Seller of the other VO, the deal is broken, and both VOs will have to go back to the market.

6 Market Repository

As stated before, markets are encapsulated as services and offer a standard API that permits the submission and withdrawal of bids. Any market mechanism can be developed by implementing the provided API. Hence, users can choose amongst a wide variety of market mechanisms when they want to trade resources. To help them in this process, DyMRA provides a market repository, where users can search markets and configurations to deploy. To deploy a service in LaCOLLA, users need to have previously uploaded the necessary files to the distributed storage service offered by the group, and to provide an specification of the service which contains, amongst other things, the files that must be retrieved for the execution of the service. The repository stores these specifications, indexed by the characteristics of the market they describe. Users can contact the repository and perform a search for some specified parameters. The repository will return a list of specifications that accomplish the requirements of the user, and these specifications can be directly used to deploy the desired market in the group.

The market repository is itself a service, deployed inside the group by an administrator. The administrator of the group must decide which market mechanisms are suitable for the group and put them in the repository in the time of deployment. He must also make sure that the files required for deploying each market mechanism are available in the group's distributed storage.

Market repositories can also be published to other groups. To do this, the publisher must advertise itself in the Market Directory, which is available to all groups. This way, users wanting to deploy a market who don't have a market repository deployed in their group, or are not satisfied by the market types offered by it, can search for a market repository in the Market Directory. Once they find a market repository, they use it to search a market type that suits their needs. If such a market type is found, the repository will send the files and the specification to the user. In order to deploy the market in her group, the user will upload the received files to the distributed storage

```
public interface marketApi {

//Registration
public RegistrationID register(Participant part);
public boolean unregister(RegistrationID rid);
public boolean hasRegistered(ParticipantID pid);

//Bidding
public BidStatus getStatus(BidID id);
public boolean withdraw(BidID id);
public BidID submitBid(BidID id, BidInfo bid);
public int getBidCount(BidderType type);
public BidID createBiddingSession(RegistrationID id);
public boolean closeBiddingSession(BidID id);

//Configuration Timouts
public void setClearingStartTime(long time);
public void setClearingEndTime(long time);

public long getClearingStartTime();
public long getClearingEndTime();

public void setWhenToClearProperty(String property);
public String getWhenToClearProperty();

//bidding times
public void setBiddingStartTime(long time);
public void setBiddingEndTime(long time);
public long getBiddingStartTime();
public long getBiddingEndTime();
//inactivity timeout
public void setBiddingInactivityTime(long time);
public long getBiddingInactivityTime();
//number of rounds
public void setNumberOfRounds(Integer round);
public Integer getNumberOfRounds();
public Integer getCurrentRound();

}
```

of her group. Because the service specification addresses the files using an identifier internal to the group, a simple modification of the specification will be needed before deployment. After this, the market will be deployed in the group.

As explained in the trading process, markets are usually created by Resource Prospectors or by Sellers, when they can't find any existing market that satisfies their requirements to trade their resources. The type of market they will deploy in these cases, as well as the location where these markets will be obtained from (inside or outside the group) is determined by the group's policies.

7 Agreement and Payment

One issue addressed in DyMRA is that of agreement and contract specification and its relation with payment. In DyMRA groups make use of virtual currency that is managed by the Accounting service. At initialization each group has a certain quantity of virtual currency that may be spent on acquiring other external resources. Besides, groups may increase their amount of currency by selling their unused resources to other groups. The following subsection present how DyMRA deals with agreements and manages payment of its transactions.

7.1 Agreement

The agreement specification process is carried out by the Market that creates an agreement object. The Agreement specifies the acquired resources, the lease time, the price and the contact point (IP address and port), that is the Seller location for the case of the Agreement sent to the Prospector and the Prospector location for the case of the Agreement sent to the Seller . Besides, the Agreement also includes the final price determined by the market's pricing policy. Once the agreement has been created, it is send to both Prospector and Seller in order to either start payment protocol or to start the leasing of resources. That choice may depend on the payment policy configured for that trade.

The agreement has been specified as an XML schema extension for Agreement, which is applied, but not limited, to standard languages for job submission like WS-Agreement[10], JDL-GLUE[11] or the JSDL [12]. JSDL and JDL provide semantics for web services description but do not address the description of Agreement. Contrarily, the WS-Agreement addresses the Agreement specification but it is placed in a more generic level than our approach which can be included as a part of it.

The following code snipped presents our XML schema extension for the agreement and contract specification.

```
<?xml version="1.0" encoding="UTF-8"?>
<xsd:schema xmlns:xs="http://www.w3.org/2001/XMLSchema" elementFormDefault="qualified">

<xsd:complexType name ="agreement Type">
 <xsd:sequence>
   <xsd:element ref ="resource"minOccurs ="1"maxOccurs ="1"/>
   <xsd:element ref ="time" minOccurs ="1"maxOccurs ="1"/>
   <xsd:element ref = "price"minOccurs ="0"maxOccurs ="1"/>
   <xsd:element ref ="partnerLocation" minOccurs ="0"maxOccurs ="1"/>
 </xsd:sequence>
</xsd:complexType>

<xsd:complexType name ="resourceType">
 <xsd:sequence>
```

```xml
      <xsd:element name="ResourceDescRef" type="xsd:IDREF"/>
   </xsd:sequence>
</xsd:complexType>

<xsd:complexType name ="leaseType">
  <xsd:sequence>
    <xsd:element name="startTime" type="timeType"/>
    <xsd:element name="endTime" type="timeTypet"/>
    <xsd:element name="slotSize" type="xsd:int"/>
    <xsd:element name="nbSlotTime" type="xsd:int"/>
  </xsd:sequence>
</xsd:complexType>

<xsd:complexType name ="locationType">
  <xsd:sequence>
    <xsd:element name="address" type="xsd:string"/>
    <xsd:element name="port" type="xsd:int"/>
  </xsd:sequence>
</xsd:complexType>

<xsd:complexType name ="timeType">
  <xsd:sequence>
    <xsd:element name="day" type="xsd:int"/>
    <xsd:element name="hour" type="xsd:int"/>
    <xsd:element name="minute" type="xsd:int"/>
    <xsd:element name="second" type="xsd:int"/>
    <xsd:element name="millis" type="xsd:int"/>
  </xsd:sequence>
</xsd:complexType>

<xsd:element name ="Agreement"type ="agreementType"/>
<xsd:element name ="resource"type ="resourceType"/>
<xsd:element name ="time" type ="leaseType"/>
<xsd:element name ="price" type ="xsd:double"/>
<xsd:element name ="partnerLocation" type ="locationType"/>

<redefine schemaLocation="./agreement.xsd">
<!-- redefinition of Agreement -->

<xsd:complexType name="pricingPolicyType">
  <xsd:choice>
    <xsd:element name="payBefore" type="xsd:boolean"/>
    <xsd:element name="payAfter" type="xsd:boolean"/>
  </xsd:choice>
</xsd:complexType>

<complexType name="Contract">
 <complexContent>
  <extension base="xs:Agreement">
    <sequence>
     <element name="currency" type="xsd:string"/>
     <element name="accountId" type="xsd:string"/>
     <element name="credentials" type="xsd:string"/>
     <element name="paymentLocation" type="locationType"/>
     <element name="accessLocation" type="locationType"/>
     <element name="paymentpolicy" type="paymentPolicyType"/>
    </sequence>
  </extension>
 </complexContent>
</complexType>
</redefine>
```

7.2 Payment

DyMRA implements two payment policies, namely "paybefore" and "payafter". "paybefore" requires the execution of the payment protocol before the Pool Service is able to use the acquired resources. "payafter" does not require the payment until the resources have already been used.

The payment protocol is executed between the Seller and the Prospector. Once the seller has received the agreement object he creates the contract object that specifies mainly the accountID of the seller, the payment service location, i.e. the accounting service within Seller's VO, the currency used for the payment (currently in DyMRA we only use one type of currency but we devised the schema to support different types of currency).

Three messages constitute the payment protocol:

1. *requestPayment*: The Seller creates the contract indicating its accountID, the location of the accounting service of the VO, the payment policy and the credentials to access the resources only for the case of a "payafter" payment policy.
2. *payment*: When the Prospector receives the contract, he determines whether it has to send the payment to the accounting service specified in the contract, or contrarily he can get access to the resource before payment. In this case, the contract includes the credentials to access to the acquired resource.
3. *paymentAcknowledgement*: When the Seller is notified by the accounting service within its VO that the payment has been done he sends the contract again with the credentials to access the resource as well as the location of either the resource or the SaleHandler responsible of that resource (only for the case of "paybefore" payment policy).

As state before, the payment protocol can be done before accessing to the resources, in this case, once the Prospector has the credentials to access the resource he communicates them to the Pool Service within its VO who will be responsible of accessing the resources. For the case of "payafter" payment policy, the Prospector communicates the Pool Service the credentials and the access point to resources. Once the lease expires, the Pool Service notifies to the Prospector that sends the *payment* to the corresponding accounting service.

8 Market Description

Semantics have an important role in our design. Buyers and Sellers have to be able to describe the main properties of the markets that they will participate in. Thus, we developed a market description language able to provide useful semantics to describe any type of market. A set of properties have been identified in order to understand the main attributes for which a market can be described.

Due to space limitations we only provide a brief description of the main classification we have done. For further details we refer readers to [13]. Besides, we are working on the complete development of an XML based language for market retrieval and description.

Registration refers to whether the market requires registration or not. Some markets may require bidders to get registered before submitting bids while others may be open to any participant.

Privacy refers to the information available by bidders at bid submission time. Using Sealed bids, bidders may simultaneously submit bids to the auctioneer without knowledge of the amount bid by other participants. Public bids provide information to other bidders in the market.

A market may be structured (Structure property) in different manners. Forward markets are initiated by one seller and are open to multiple buyers. Reverse market are in contrast initiated by a buyer requiring offers from several sellers. Finally double markets enable multiple buyers to trade with multiple sellers. The Auction property is used to describe the main characteristics of the mechanism used to allocate resources. For example, whether it accepts withdraw of bids or not, the type of items traded in the market, the type of clearing, the type of feedback amongst others. Other interesting properties are those related with technical aspects of the mechanism such as the provided economical efficiency (optimal or not) or the size of bids that the mechanism is able to deal with. Finally other properties such as Quality of Service (QoS) allow the specification of upper bounds on time for the winner determination computation.

9 Markets in DyMRA

In this section we focus on the type of markets that can be deployed in DyMRA. As stated before markets are encapsulated as services and offer a standard API that permits the submission and withdrawal of bids. Any market mechanism can be developed by implementing the provided API. As our understanding, this constitutes an added value to DyMRA since it provides support to any type of market. For testing purposes we developed two different auctions, namely the Continuous Double Auction (CDA) and the Scheduled Double Auction (SDA) that aimed to: a) demonstrate the suitability of the framework to support different market mechanisms. b) understand the properties of each of the mechanisms when they are used to trade computational resources.

9.1 The Double Auction

A widely used mechanism to allocate computational resources is the Double Auction (DA)[14]. The classical DA clears multiple units of single items amongst multiple buyers and multiple sellers. It does not guarantee complete satisfaction when a participant submits bids for several items. We can find different types of DA; The Scheduled Double Auction (SDA) is the key example of a discrete-time double-sided auction. Its key feature is that bids and asks are collected over specified intervals of time and then "cleared" at the expiration of the bidding interval. Given the supply and demand revealed in the bid, a market-clearing price is determined to maximize market efficiency. The Continuous Double Auction (CDA) is a continuous-time double-sided auction, i.e. there is no clearing-time frame. Bids and asks are continuously received and matched. Trades can occur at any time; hence, there is continuous matching and clearance. The trades consist of bilateral transactions triggered by an acceptance of the best bid or ask. The matching bid and ask will be removed from the auction to form a transaction. Many such individual transactions are carried out and trading does not stop as transactions are concluded.

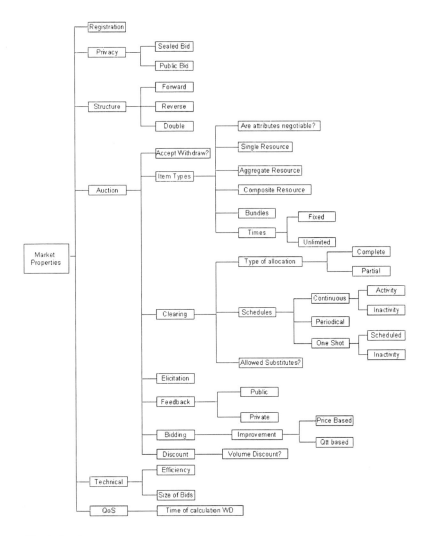

Fig. 4. Service taxonomy

10 Market Evaluation

This section presents the evaluation and comparison of the continuous-time double auction and the discrete-time double auction.

10.1 Test Configuration

In order to obtain data from the two market institutions, two market instances were activated. The first market instance implemented a continuous double auction

(CDA), and the second market instance implemented a scheduled double auction (SDA). Tests were conducted during 500 iterations. In each iteration, 100 processes trying to buy resources and 100 trying to sell resources were executed. 50 of the buying (selling) processes submitted bids to the CDA and the rest submitted bids to the SDA. Processes acting as buyers generated bids following a Pareto distribution having established an initial price. Pareto distribution has been considered the most appropriate distribution for generating bids since it has been used to describe the distribution of wealth in the society. Sellers generated their offers following a normal distribution, because of the normal distribution of the costs of the traded items. Prices were updated after each iteration taking into account the variations in supply and demand.

10.2 Results

Two important variables were observed, the number of allocations and the social welfare. The first one indicates the efficiency of the mechanism in terms of the number of resources used. Social welfare measures the capacity of the market to allocate the resources to those who need them more. Social welfare has been calculated as the aggregation of the buyers surplus and sellers surplus.

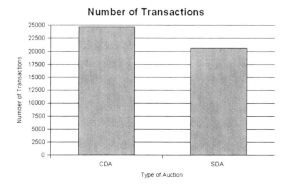

Fig. 5. Number of successful transactions for both mechanisms

Figure 5 presents the number of successful transactions for both mechanisms. The CDA obtained a higher number of transactions than the SDA, this is of course because, each time a bid meets an ask with a price lower than its price a transaction occurs. Contrarily, figure 6 presents the social welfare for the 500 iterations. In our tests, the CDA obtained a social welfare of 239036,29 euros while the SDA obtained 245202,44 euros with 4101 less transactions. This happens because the SDA calculates the price *(p)* where supply balances demand and matches the highest bids above the price *(p)* with the lowest asks below the price *(p)*. Figure 7 shows the average price evolution in the market as well as the social welfare for each of the

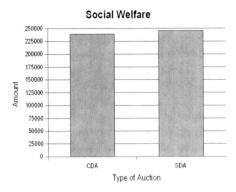

Fig. 6. Social welfare

iterations. Price evolution is guided by random variables that depend on the levels of activity in the market.

Our tests showed that the social welfare provided by the SDA is higher than the social welfare provided by the CDA. Contrarily, the CDA provides a higher number of allocations which means that a higher number of resources are utilized. The choice of one or the other constitutes a trade-off between economic efficiency and number of allocations. Which mechanism to use is an individual decision that the trader may take based on its strategy.

After doing the presented tests, we wondered about improving the social welfare provided by the CDA without decreasing significantly the number of allocations. To do so, we delayed the immediate allocation of the CDA and re-executed the tests. In order to delay the clearing of the CDA it was configured to wait until at least 4

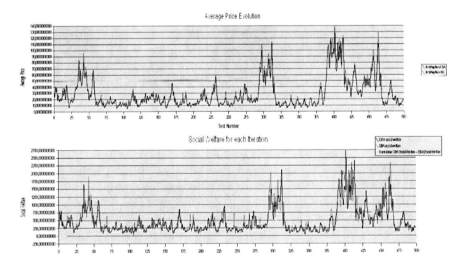

Fig. 7. Average price evolution and social welfare

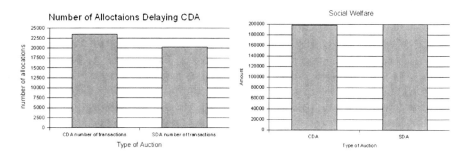

Fig. 8. Number of allocations provided by each algorithm

bids had arrived to the market. The results were fairly interesting. As can be seen in figure 8, the number of allocations provided by the CDA remained higher than the number of allocations that the SDA provided, contrarily the difference of the aggregated social welfare provided for both methods was reduced nearly a 50%.

By delaying the clearing of the CDA the number of allocations is slightly reduced at expenses of increasing the social welfare. However, the number of allocations remains higher than the number of allocations provided by the SDA and the improvement of the social welfare is more significant than the decrement of the number of allocations. A conclusion extracted is that the efficiency of the CDA can be improved by delaying slightly the clearing process. It can be done either by discretizing continuous time into small iterative time frames or by delaying the clearing until a certain number of bids arrive to the market.

11 Validation of DyMRA Mechanisms

This section presents an implementation of DyMRA and its validation. These results demonstrate the viability of our proposal and encourage us to refine it. Currently we are working on a further and exhaustive validation.

We implemented a prototype of the proposed architecture to test its usefulness. The Prospector, Seller, Pool, SaleHandler and the Market have been implemented as deployable services over the LaCOLLA middleware. The Market provides generic operations that allow different mechanisms to be implemented. For our testing purposes we developed a double auction [15] protocol that enables buyers and sellers to submit bids for multiple units of a single resource (i.e storage capacity, cpu capacity and applications).

The MarketDirectory has been implemented as a centralized index, but, as mentioned above, it can be easily substituted with a decentralized approach [16, 17]. For our testing purposes, the market directory stores pairs of $< key, value >$ where the key identifies the type of traded resource and the value refers to the identifier of the market where it is traded in.

The objective of our test is to validate the trading process described above. One of the main objectives of our proposal is to provide good availability in environments of high dynamism and churn. Hence, availability has been the main focus of our tests.

We executed a process which periodically tried to buy resources, and another that tried to sell resources. The necessary services (Prospector, Pool, Seller) where active inside the VO, while there was a MarketDirectory available in a static location. Markets, though, according to our proposal, are created on demand. When a Prospector or a Seller wants to access a Market, but there in't any available, it proceeds to create and activate one. When this happens, it is counted in our tests as a failed attempt. For simplicity, Markets have been assigned a limited lifespan, after which they resolve the auction and send the results to the clients. This implies that, periodically, a Prospector or a Seller will have to create a Market, thus decreasing the perceived availability. Markets could also be permanently active, which would increase the availability of the system. There is, though, a trade off between the obtained availability and the resources spent to keep the market active.

The LaCOLLA middleware offers the ability to simulate users' activity and system dynamism (connections, disconnections, failures) in order to conduct tests and validate its functioning. We measured the availability of markets in function of the levels of dynamism of the system. Specifically, we evaluated two different levels of dynamism. In the less dynamic (from now on, called G1) each component had a probability of failure per iteration of 0,0005, and a probability of ordered disconnection of 0,0025. In the more dynamic of the two (G2), the probability of failure per iteration was 0,005, while the probability of disconnection was 0,008. Tests lasted 500 iterations.

The data we analyze is the number of bids that arrive to the market, in relation to the number of bids issued by the group. This depends exclusively of the mechanisms of our system, in contrast to the number of matches, which depends on supply and demand. Note once again that this number decreases because markets have a limited lifespan and are created on demand, which results in a failed access when a market

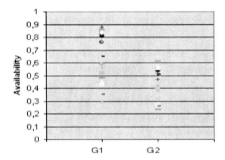

Fig. 9. Availability vs level of Dynamism

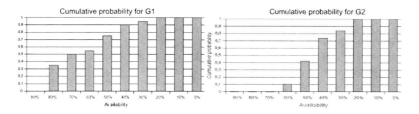

Fig. 10. Cumulative probability of availability levels for G1 and G2

must be created. That doesn't mean that, in a real situation, the bid cannot be issued, only that it will have a bigger delay.

Fig. 9 shows the availability (percentage of succesfully issued bids) obtained in 20 executions, for both G1 and G2. We see that, as expected, the availability is higher in G1, decreasing in G2 because of the higher level of dynamism.

Fig. 10 shows the cumulative distribution function for both G1 and G2. For G1, 50% of the executions obtain an availability of 70% or higher, which must be considered noting that markets are activated on demand, and we count it as unavailable when activation is needed. For G2, availability is low because of the high level of dynamism.

12 Conclusions

The chapter describes DyMRA, a framework for inter-VO resource allocation that has been specifically designed to support contributory environments such as those related with collaborative learning. The key aspect of DyMRA is that of market decentralization, that allows allocations of resources amongst different VO in spite of markets' failures. Markets and mediator components such as buyer agents and seller agents are exposed as mobile services within the VO that allows the utilization of inherently centralized mechanisms such as auctions into a decentralized environment without introducing bottlenecks or single points of failure. The paper presents the complete architecture of DyMRA as well as the main tools that has been developed to make DyMRA a reality. Semantics have been used to provide comprehensive languages to describe resources and markets. A common understandig is required for components in such a decentralized environment in order to achieve self-managing and adaptative components. The paper shows how the economy can be used to regulate the usage of resources as well as the implications of using different market mechanisms. Finally the paper presents the results of evaluating our proposed architecture. Our future work includes the complete development of the DyMRA components, such as a decentralized Market Directory, and the set of mechanisms to control the access to external allocated resources. Besides, we aim to consider duration of the allocations of resources (lease times) that would permit the application of our framework in a real environment.

References

1. Shneidman, J., Ng, C., Parkes, D.C., AuYoung, A., Snoeren, A.C., Vahdat, A., Chun, B.: Why markets could (but don't currently) solve resource allocation problems in systems. In: HOTOS 2005: Proceedings of the 10th conference on Hot Topics in Operating Systems, p. 7. USENIX Association, Berkeley (2005)
2. Marquès, J.M., Vilajosana, X., Daradoumis, T., Navarro, L.: IEEE Internet Computing 11(2), p. 56 (2007),
   http://doi.ieeecomputersociety.org/10.1109/MIC.2007.43
3. Lázaro, D., Marquès, J.M., Jorba, J.: Decentralized Service Deployment for Collaborative Environments. In: Proceedings of the 1st International Conference on Complex, Intelligent and Software-Intensive Systems, CISIS 2007, pp. 229–234. IEEE Computer Society, Los Alamitos (2007),
   http://doi.ieeecomputersociety.org/10.1109/CISIS.2007.18
4. Kevin Lai, B.A.H., Fine, L.: Tycoon: A Distributed Market-based Resource Allocation System. Tech. Rep. arXiv:cs.DC/0404013, HP Labs, Palo Alto, CA, USA (2004)
5. Catnets Consortium. Deliverable d3.1: Implementation of additional services for the economic enhanced platforms in grid/p2p platform: Preparation of the concepts and mechanisms for implementation (gmm) (2005),
   http://sorma.fzi.de/protected/Cat06.pdf
6. Buyya, R., Abramson, D., Venugopal, S.: The grid economy (2004),
   citeseer.ist.psu.edu/buyya05grid.html
7. Amara-Hachmi, N., Vilajosana, X., Krishnaswamy, R., Navarro, L., Marques, J.M.: Towards an open Grid marketplace framework for resources trade. In: GADA Conference: Proceedings of the OTM Federated Conferences and Workshops. Springer, Heidelberg (2007), http://www.cs.rmit.edu.au/fedconf/
8. Gavalda, C.P., Lopez, P.G., Andreu, R.M.: Deploying Wide-Area Applications Is a Snap. IEEE Internet Computing 11(2), 72–79 (2007),
   http://dx.doi.org/10.1109/MIC.2007.31
9. Adam, C., Stadler, R. (2006)
10. Andrieux, A., Czajkowski, K., Dan, A., Keahey, K., Ludwig, H., Nakata, T., Pruyne, J., Rofrano, J., Tuecke, S., Xu, M.: Web services agreement specification (ws-agreement), version 2006-09-07. Tech. rep., Global Grid Forum (2006)
11. Andreozzi, S., Burke, S., Field, L., Fisher, S., Konya, B., Mambelli, M., Schopf, J.M., Viljoen, M., Wilson, A.: GLUE Schema Specification - Version 1.2 (2005),
    http://glueschema.forge.cnaf.infn.it/Spec/V12
12. Anjomshoaa, A., Brisard, F., Drescher, M., Fellows, D., Ly, A., McGough, S., Pulsipher, D., Savva, A.: Job submission description language (jsdl) specification, version 1.0. Tech. rep., Global Grid Forum (2005)
13. Grid4all consortium. wp2. deliverable month 12: D2.1 requirements for grid4all virtual organisations and resource management and state of the art analysis
14. Wurman, P.R., Walsh, W.E., Wellman, M.P.: Flexible double auctions for electronic commerce: Theory and implementation. Decision Support Systems 24, 17–27 (1998)
15. Bao, S., Wurman, P.R.: A comparison of two algorithms for multi-unit k-double auctions. In: ICEC 2003: Proceedings of the 5th international conference on Electronic commerce, pp. 47–52. ACM Press, New York (2003),
    http://doi.acm.org/10.1145/948005.948012

16. Ghodsi, A.: Distributed k-ary System: Algorithms for distributed hash tables. PhD dissertation, KTH—Royal Institute of Technology, Stockholm, Sweden (2006)
17. Castro, M., Druschel, P., Kermarrec, A.-M., Rowstron, A.: One ring to rule them all: service discovery and binding in structured peer-to-peer overlay networks. In: EW10: Proceedings of the 10th workshop on ACM SIGOPS European workshop: beyond the PC, pp. 140–145. ACM Press, New York (2002),
http://doi.acm.org/10.1145/1133373.1133399

A Semantic Description Model for the Development and Evaluation of Grid-Based, Innovative, Ubiquous and Pervasive Collaborative Learning Scenarios

Gustavo Gutiérrez-Carreón and Josep Jorba

Open University of Catalonia, Av. Tibidabo 39-43 - 08035 Barcelona, Spain
{ggutierrezc, jjorbae}@uoc.edu

Abstract. The great success and demand of software tools in education have generated new challenges to improve the functionality and to reduce the limitations of time and availability of bandwidth to process the information. There are a lot of efforts to use distributed technologies in education, in particular Grid computing. The Learning Grid is based on a secure, flexible and coordinated form of sharing network resources which are dynamically collected by individuals and institutions, and establishing mechanisms for the correct exchange of information and a strict control of the resources to share. Learning services are fundamental components of learning Grid representing functionalities that can be easily reused without knowing the details of how services have been implemented. Semantic modeling of web services promises to automate tasks such as discovery, matching, composition and invocation of services. The objective of this chapter is to present an overview of a work related with the analysis, design and implementation of semantic models for the description of learning services and their incorporation inside Collaborative Learning Scenarios based on Grid technologies.

1 Introduction

The electronic learning (e-learning) arises like a promising option for people that want to study on-line. Although in the last years there is a substantial increase in the use of computers and networks, mainly as a consequence of faster hardware and more sophisticated software, there are still problems in the fields of integrating several resources that consolidate the e-learning. The education systems based on services open new forms of constructing e-learning systems using technologies that Grid computing provides, which include as their primary requirements the provision of shared services, syndicating of heterogeneous resources and taking advantage of the discovery of pertinent content. In that sense, the use of web services in general and Grid services in particular, has fundamentally changed the way to develop e-learning applications. In the field of the Learning Grid services, on the one hand an important objective to achieve is the correct integration of heterogeneous learning services offered by different organizations in order to develop collaborative learning tools, on a user-centered context. On the other hand, if a single Grid learning service cannot satisfy the functionality required by the user, one should have the possibility to combine existent services to fulfill the requirement. There are a lot of efforts to construct methods exploring how semantic technologies (capabilities) can be used to support the improvement of e-learning in general and the dynamism and personalization in particular and to define composition models for personalized Collaborative Learning Scenarios based on Grid technologies and focused on contextualized and user-centered approaches of collaborative learning, and whose main objective is the construction of active knowledge, as well as the possibility to compose and to automatically discover high level tools and learning services based on low level ones for discovery, search, matching and composition of Grid and web services that use the semantic description of capabilities as a main tool.

The rest of the paper is organized as follows: In section 2 we review the state of the art of some technological efforts to construct Learning Grid scenarios. Section 3 presents a vision of the Learning Grid Architecture, some approaches to define semantics models of learning grid scenarios and a proposal of personalized Learning Grid scenario based on Portals. Finally in Section 4, we present the conclusions of this work.

2 Related Efforts to Construct the Learning Grid

The characteristics of technologies applied to learning Grid scenarios will mean a great advantage for learning process, chiefly to increase the efficiency of learning for individuals and groups and to contribute to a deeper understanding of the learning process by exploring links between human learning, cognition, and technologies. Below we mention the most recent and significant works that contributed to the development of technologies that make possible the vision of the Learning Grid.

The Grid Service Based Portal for Virtual Learning Campus [1] developed an environment which makes use of the Grid capabilities so that to make possible the dynamic sharing and coordination of heterogeneous resources which are found dispersed in the network. The project focuses on the development of a video digital library based on Grid for a Virtual Campus that allows an easy access and implementation of several services. In spite of being a project that aimed to take advantage of the capabilities that Grid technology provides, it is limited on a unique type of educative resources, like video, which a structure of services is developed for.

Methodology for Supporting Novel Model of E-Learning Platform in Grid Architecture [2] presents a combination of grid technology with E-Learning, present an E-Learning architecture. They propose a mechanism to integrate unrelated computers in schools to replace high level server as the teaching platforms of E-Learning for saving cost and time.

In [3] the authors describe a platform of e-learning based on Grid service technologies. In this platform the supply of virtual learning services designated for students, instructors and course suppliers is based on the resource administration for group collaboration based on Grid, allowing ubiquous access to information and taking advantage of the potentiality of the computer systems. On the one hand, the advantage of this proposal is that it is the first one that elaborates on the use of Grid resources and their description through Grid and web services technologies, in particular WSDL. On the other hand, it dictates the need for the development of a semantic model description that enables a more complete description of learning resources.

ULabGrid, an Infrastructure to Develop Distant Laboratories for Undergraduate Students over a Grid, [4] proposes a new architecture that allows the educators to design remote collaborative laboratories for university students using the Grid infrastructure. This project is one of the first in its type in trying to combine the facilities that Grid provides in a practical scenario in order to achieve resource sharing and motivate collaborative work. In this sense the design of Grid-based collaborativeLearning Scenarios should be supported by semantic descriptions that allow the best tracking of resources available in the network.

Another work that aims at developing a Generic E-learning Engineering Framework Embracing the Semantic Web [5] proposes the convergence of e-learning, Web semantics and e-business by introducing a generic engineering approach that labels learning objects with RDF for semantic e-learning and integrating it with a process oriented paradigm. This work can serve as one of the first approaches as regards the use of information modeling and RDF to label Collaborative Learning Scenarios resources, making use of a process management approach, which if adapted to the Learning Grid will provide a new generation of applications.

A further work proposes an Agent-Based Robust Collaborative Virtual Environment for E-Learning in the Service Grid [6]. In this virtual environment, all Web resources and services are accessed via service encapsulation, which may result in a more scalable and robust collaborative learning architecture.

In [7], the authors present a Semantic Grid for E-leaning based on DartGrid, and also put forward a dynamic, extensible Semantic-based distributed infrastructure for E-Collaborative Learning Scenarios. They explore the essential and fundamental roles played by RDF semantics for Elearning resource sharing, and implement a set of semantically enabled tools and grid services for E-learning such as semantic browser, ontology service, semantic query service, and semantic registration service.

KGCL, a Knowledge-Grid-Based Cooperative Learning Environment [8], supports the cooperation between a person and the computer at a knowledge level, and allows the enrichment not only of the resources in the Knowledge Grid but also of the users' knowledge by means of knowledge refinement, knowledge reuse and the online meeting of participants. The KGCL prototype has been currently applied and is available for online use. Experiments have shown that the environment can promote the effectiveness of group work. This system has also shown the great impact that Grid technologies can have even though no model of semantic description was implemented that could improve its performance.

In [9], a Grid Service Framework for Metadata Management in Self-e-Learning Networks focuses on how the use of metadata can be critical for Grid systems. More specifically, the semantic description constitutes a very beneficial extension of Grid environments. The Self e-Learning Network (Se-LeNe) is used as a test application while a set of services is proposed which are implemented with OGSA [10]. The project focuses on providing services that use learning objects metadata, based on a sufficiently generic approach so that they can be used by other Grid-based systems which need to make use of semantic descriptions.

The Semantic Grid for human learning is one of the main objectives of the integrated project ELeGI (European Learning Grid Infrastructure) [11]. The objective of ELeGI is to provide an advance in the current practices of learning through the definition and implementation of a software architecture that achieves to unify the semantic Grid and information technologies in order to promote and give support to the definition and adoption of learning paradigms for the construction of knowledge that combines customized and ubiquous techniques based on experiential, collaborative and contextualized learning.

In this line, [12] presents a work about ontology based user modeling for personalization of Grid learning services. This work describes how the learning services of the semantic Grid should support a user-centered, customized, contextualized, experiential and ubiquous based learning approach. They claim that in order to provide a customized learning process, it is necessary to study and define methodologies that represent the context of learning and student through suitable knowledge structures, such as the ontologies. This work focuses then on the role that customized ontologies may play on a new generation of intelligent services; more specifically, it explores the role of ontologies to obtain Gridbased learning services in ELeGI.

PLANT [13] is a distributed architecture for personalized E-Learning built upon the Edutella network which is a schema-based peer to-peer system. The main objective of PLANT is to facilitate individual learning on the Internet which abounds in a wide variety of educational resources and services. PLANT allows users to conduct complex queries for best results according to their knowledge backgrounds and learning goals. With the distributed resource evaluation algorithm based on consensus, the quality of education resources can be precisely estimated, which stimulates resources to evolve in the network. By providing a rich set of learning assistant services, individual users can get good support to achieve their learning goals.

The SELF project [14] proposes a learning environment that results from the integration of several technologies, specially the semantic Web, Grid technology, collaborative tools as well as customized tools and knowledge management techniques. SELF provides a mechanism for the intelligent search of services making use of semantic description tools. This project presents an important reference of the use of different technologies for the development of Grid-based Collaborative Learning Scenarios, even though it is not based on semantic description models for the definition of its tools.

OntoEdu [15] is a flexible platform for online learning which is based on diverse technologies like ubiquous computing, ontology engineering, Web semantics and computational Grid. It is compound of five parts: user adaptation, automatic composition, educative ontologies, a module of services and a module of contents; among these parts the educative ontology is the main one. The main objectives of OntoEdu are to obtain reusability of concepts, adaptability for users and devices, automatic composition, as well as scalability in functionality and performance. In the near future, this platform aims to be adapted to a Grid environment so that it can carry out its activities based on distributed computing.

The work developed in [16] presents a workflow framework for pervasive learning objects composition by employing a Grid services flow language. The learning objects are distributed in heterogeneous environments which have been used in order to allow effective collaboration and the reuse of learning objects; this fact can help users to learn with no limitations of time and space. This work shows the great opportunities that exist in those research groups which make use of Grid technology to develop innovative, pervasive and ubiquous Collaborative Learning Scenarios. Though this research work is still encountered at an initial phase, it can be further enhanced by the application of semantic description of learning services.

In the work described as "Extending and Augmenting Scientific Enquiry trough Pervasive Learning Environments" [17], the authors show the advantages of using pervasive Collaborative Learning Scenarios that take advantage of mobile and wireless devices. This fact is used to integrate several scientific processes and provide support to students that carry out activities outside the class.

Finally, in the work referenced as "Semantic Search of Learning Services in a Grid- Based Collaborative System" [18], the authors have constructed an ontological description for collaborative work tools that allow one to make a manual search of the diverse resources that these tools provide within a Grid environment with the minimum of technical knowledge. This work proposes a Grid-based tool called Gridcole, which can serve as a basis to implement different conceptual approaches of Grid-based semantic description of learning services, thus extending and endowing it with an innovative, pervasive and ubiquous projection.

3 Defining Collaborative Learning Grid Scenarios with Semantic Capabilities

In [19] the authors introduce the concept of Semantic Grid and they define it as an extension of the current Grid in which information and services are given well-defined meaning, better enabling computers and people to work in cooperation. They highlight how higher-level knowledge plays an important role in future Grid applications, issues related to the representation, discovery, and integration of domain knowledge, data mining, knowledge discovery, text and multimedia content analysis, semantic information extraction and integration, etc are relevant. They distinguish between a Knowledge Grid which is a Grid of semantics based on knowledge generated by applications using and mining the Grid and a Semantic Grid which manages semantics for the Grid to manage and execute its architectural components (Figure 1). In this section we analyze the architecture of learning Grid and based on this we review some approximations to how describe it semantically. After this, we make an evolution of models of semantic representation looking for a better way of implementation. To conclude the section we

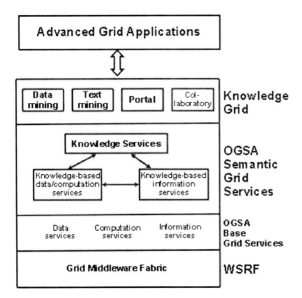

Fig. 1. The Semantic/Knowledge Grid Stack

will introduce a proposal of personalized learning scenario that promotes the collaboration, the ubiquitous and pervasive access and that is based on the use of portals.

3.1 Learning Grid Architecture

In [20] the authors describe the paradigm of learning based on Grid as follows:

A Learning Grid is an enabling architecture based on three pillars: Grid, Semantics and Educational Modelling allowing the definition and the execution of learning experiences obtained as cooperation and composition of distributed heterogeneous actors, resources and services.

According to this, they define a layered architecture for the Learning Grid and explain how it can be used as a basis to provide Learning Services and Applications (Figure 2):

1. The Infrastructure Services provide an implementation of the Web Service Resource Framework specifications in order to define the underlying service model;
2. the Grid Middleware for VO Management provides an implementation of the services identified by the Open Grid Services Architecture allowing to create and manage a distributed VO and integrate, virtualize and manage resources and services upon it;
3. the Semantic Annotation, Discovery and Composition Services provide learning independent functionalities based on specifications and languages for the semantic description of Web services to allow the automatic negotiation, discovery and composition of Grid Services;

4. the Educational Modeling and Execution Services provide contextualized features related to the formal description of learning experiences basing on Education Modeling Languages and the automatic discovery, composition and execution of learning resources and services available on the Grid basing on such descriptions.

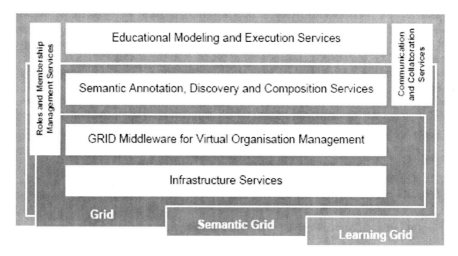

Fig. 2. Architecture of a Learning Grid [20]

A Grid is composed by Infrastructure Services plus a Grid Middleware for VO Management; a Semantic Grid is composed by a Grid plus Semantic Annotation, Discovery and Composition Services; a Learning Grid is composed by a Semantic Grid plus Educational Modelling and Execution Services plus a set of "environment" services described below to support the creation, the operation, the evolution and the maintenance of a learning community:

- the Role and Membership Management Services manage users, groups, roles and membership inside VOs on the Learning Grid by supporting the Grid Middleware for VO Management as well as Semantic Annotation, Discovery and Composition Services and Educational Modelling and Execution Services;
- the Communication and Collaboration Services provide tools to support communication and collaboration among participants in groups, communities and actors involved in learning experiences by supporting Semantic Annotation, Discovery and Composition Services and Educational Modelling and Execution Services.

3.2 Semantic Modeling of Collaborative Learning Scenarios

Using ontology technologies such as RDF, DAML+OIL, OWL-S and WSMO [21, 22], it is possible to create semantically rich data models that are denominated semantic schemas. These semantic schemas are made up of triples (subject-predicate-object), where subjects and objects are entities, and predicates indicate relationships between those entities. Users can define their own properties, as well as their own classes. Instances of these classes can then be created and described with values for related properties. In these schemas there is more implicit information than it can be usually found in their text representation. Each triple forms a graph with two nodes connected by an edge. Each instance can have several properties, and that graph can be expanded to have many nodes connected to the central instance. Finally, when two instances are connected via a property, their respective sub-graphs become connected.

The usage of an ontology is of interest whenever the costs that arise through terminological disagreements and misunderstandings while not using ontologies exceed the costs for providing ontologies and formalized descriptions of situations [22]. There are a number of characteristics of settings where use of ontologies appears promising:

1. Important heterogeneous (and possibly imprecise) vocabularies
2. Small to medium sized domain-
3. Multitude of participants with overlapping interests.
4. Long-term interest in understanding of vocabulary and corresponding data
5. Many and/or (rather) expensive transactions

The definition of semantic modeling is a work that began with the election of an e-learning framework. We selected the IMS Global Learning Consortium Abstract framework [23], which represents a set of services used to construct an e-learning system in its broadest sense. One of the design principles for the abstract framework is the adoption of a service abstraction to describe the appropriate e-Learning functionality (Figure 3).

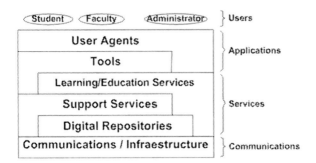

Fig. 3. A logical architecture for an eLearning system

There are some works related to the semantic description of Grid Learning Services. OntoEdu [15] is compound of five parts: user adaptation, automatic composition, educative ontologies, a module of services and a module of contents; among these parts the educative ontology is the main one. The main objectives of OntoEdu are to obtain reusability of concepts, adaptability for users and devices, automatic composition, as well as scalability in functionality and performance. In the near future, this platform aims to be adapted to a Grid environment so that it can carry out its activities based on distributed computing (Figure 4).

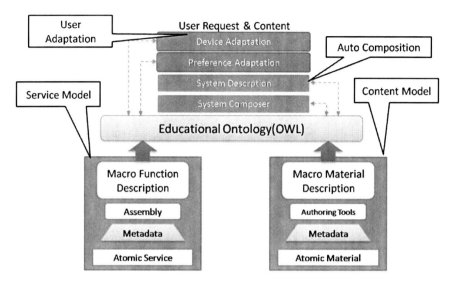

Fig. 4. A logical architecture of OntoEdu

In Gridcole [18], the authors said about their proposed information model of the ontology of learning tools (Figure 5): "...*the ontology should allow to describe what types of activities does a particular tool support, either individual or collaborative. This issue is expressed independently of the type of learning tool, as shown pictorially in Figure...*".

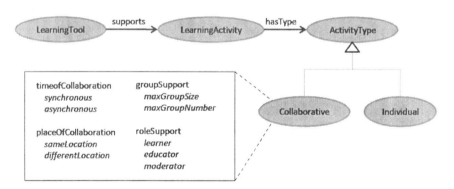

Fig. 5. Information model of the ontology of learning tools

Another effort to construct a semantic model is reported on [24] and it's based on Semantic Markup for Web Services (OWL-S, formerly DAML-S)[25]. OWL-S semantically rich descriptions enable automated machine reasoning over service and domain descriptions, thus supporting the automation of service discovery, composition, execution, and reducing manual configuration and programming efforts.

OWL-S is motivated by the need to provide three essential types of knowledge about a service [25] (Figure 6): the *Service Profile* describes what the service does by specifying the input and output types, preconditions and effects, the *Service Model* describes how the service works, the *Service Grounding* contains the details of how an agent can access a service by specifying a communications protocol, parameters to be used in the protocol and the serialization techniques to be employed for the communication.

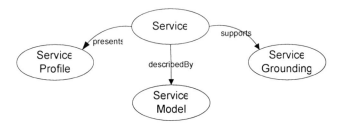

Fig. 6. The top level of the service ontology

The model proposes a mechanism using OWL-S to describe semantically a Grid Based Learning Service (GBLS). One the one hand, we identify the principle characteristics of a GBLS related to an e-learning environment and the activities that support it.

On the other hand, we consider a GBLS as a granular functional component with some input information, a functional activity and some output information. In this sense we form two conceptual groups of properties to achieve a complete description of a GBLS: the Learning Services Identification (LSI) and the Learning Services Access Point (LSAP) (Figure 7).

The LSI element constitutes a set of parameters used to define the principle characteristics of a GBLS. Table 1 describes it in more detail.

The LSAP element is characterized by the most important functional properties of a GBLS and is described in Table 2.

Fig. 7. Parts of the conceptual model for a GBLS semantic description

On the one hand, the LSI contains basic information related to a learning service, allowing a user centric search. This model is generic enough to be implemented in any e-learning framework supported by GBLS and can be used for describing both low-level and composed services. The domain of categories could be adapted to any ontology or taxonomy of services. For example, if we adopt the IMS Abstract Framework Service's categories, we can obtain a taxonomy of learning tools supported by Common Services and they are in turn supported by Basic Services (Figure 8). On the other hand, LSAP allows the construction of an

A Semantic Description Model

Table 1. LSI Elements

Learning Services Identification			
Element	Description	OWL's Element	Parameter
Name	The name of a service	Service Profile	ServiceName
Category	Depending on the e-learning framework, services providers could construct the domain of categories for each group of learning services	Service Profile	ServiceCategory
Description	General description of a service	Service Profile	textDescription

Table 2. Properties and resulting parameters of LSAP

Learning Services Access Point			
Element	Description	OWL's Element	Parameter
Activity	Activity in the e-learning framework supported by a service	Service Model	hasParameter
User	Defines the profile that makes use of a service (learner, teacher, another process, etc)	Service Model	hasParameter
Related Services	Specifies one or more services related to a service	Service Model	hasParticipant
Process	A service is described as a functional process	Service Model	Pocess
Errors	Specifies the errors resulting from the execution of a service	Service Model	hasParameter
Bindings	Definitions included in a WSDL description of a learning service	Service Grounding	WsdlAtomicProcessGrounding

ontology domain for functional parameters related to the e-learning framework. This semantic description, in combination with the modifications suggested in [26], allows capability-based search as well as discovery of learning services based on the inputs and preconditions that need to be satisfied and on the outputs and effects that need to be produced. The Bindings element of LSAP is a parameter of WSDL describing the interface of each learning service. Both elements of the model are necessary to deal with the problem of using and integrating low-level services to compose more complex high-level services or tools. In particular, the LSAP describes grid-based learning services as processes, which allows one to specify whether a service is an atomic, simple o composite process as well as its relationship with the other services.

Fig. 8. Service Categories for the IMS Abstract Framework

3.3 Evolution of OWL-S to WSMO

The authors of [27] comments that of the models for semantically annotating Web Services, WSMO [28] (Web Service Modeling Ontology) and OWL-S are the most closely related. Both aim at the provision of a comprehensive conceptual model for Semantic Web Services. WSMO describe how an important foundation point of the work on WSMO was the mode provided by OWL-S but maintain that OWL-S has a number of serious fundamental flaws that give rise to problems when attempting to use the ontology in practice. In WSMO the viewpoint of service requester and service provider are distinctly represented by the complementary concepts of goals and Web Services. This separation is adopted from the research in the problem-solving domain and is a clear point of distinction between the OWL-S and WSMO models.

WSMO provides a conceptual model for the description of Web services. WSMO distinguishes between user goals, which are descriptions of the desires of the requester, and Web service descriptions, which are descriptions of the functionality and interface of the service offered by the provider. This is one of the distinctions between OWL-S and WSMO. In OWL-S the service concept is used to describe both services and requests for services. Although from a modeling viewpoint WSMO goals and Web Services contain the same structure, they represent different perspectives in the conceptual model and for this reason are kept separate. Figure 9 presents the elements of a Web service description, namely non-functional properties, a capability, a choreography and an orchestration. The term interface is used to describe the combination of the choreography and orchestration of a service.

WSMO makes a distinction between the inputs and outputs of the service, and the state of the world. Based on these considerations a capability description

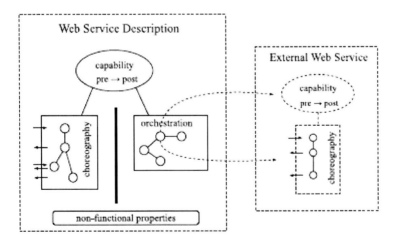

Fig. 9. Elements of WSMO Web service description

comprises four main elements. Preconditions describe conditions on the state of the information space prior to execution. There may exist additional conditions that must hold in the real world in order for the service to successfully execute. These conditions, called Assumptions, are not necessarily checked by the service before execution but are crucial to the successful execution of the service. Postconditions describe conditions on the state of the information space after execution has occurred, thus describing properties of the outputs of the service, as well as the relationship between the inputs and the outputs. Effects are conditions that are guaranteed to hold in the real world after execution. The interface of a Web service specifies how to interact with the service in terms of a choreography, this choreography essentially provides information about the relationships between different operations on the Web service. The interface of a Web service description also contains an orchestration description. An orchestration specifies which services this service relies upon to provide its functionality.

3.4 Grid Learning Portals: Personalized, Ubiquous and Pervasive Collaborative Learning Scenarios

In [29] the authors comment that ubiquitous computing environments are different from what one traditionally finds in most school settings. It offers to all students and teachers continuous access to a wide range of software, electronic documents, the Internet, and other digital resources for teaching and learning. These initiatives' goals include increasing economic competitiveness, reducing inequities in access to computers and information between students from wealthy and poor families, and raising student achievement through specific interventions.

Grid portals have emerged as a solution for the end users who want to have access, on the one hand, to a diverse set of heterogeneous resources of computational grids, and on the other hand, to a web portal joining in the same point information from diverse sources in a unified way, making use of a wide variety of technologies and services.

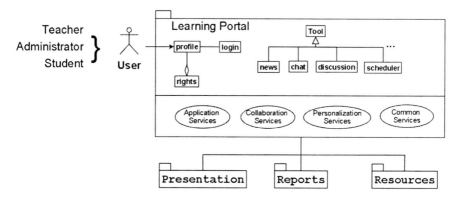

Fig. 10. Interaction between users and Learning Portal

In particular, learning portals allow people to interact and work together as a distributed team. Groups can dynamically form on projects or sub-projects or look for purposes requiring specific tools and authorization. The proposal we present is based on Grid portals in order to construct personalized Collaborative Learning Scenarios. Figure 10 shows the interaction between a user and the tools, services and resources provided for this kind of scenarios. In that sense the proposed model of Learning Grid portal incorporates a semantic engine (Figure 11) that allows the interaction among diverse levels of the structure of a learning framework based on Grid.

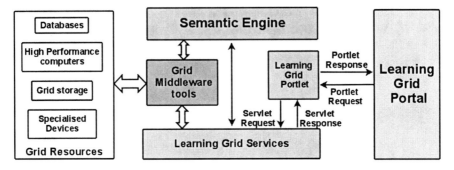

Fig. 11. Learning Grid Portal model with semantic capabilities

The objective of this model is to have containers of learning services that make use of diverse technologies based on grid and at the same time they can be invoked by a Portlet and presented to the end user. The semantic capabilities of this model will allow the automatic discovery, search, composition and invocation of services with direct benefits for end users. For the semantic representation of the model we use the Web Service Modeling Language (WSML) [30], which provides means for describing the functionality and behavior of learning services, as well as the underlying educational knowledge in the form of ontologies, with a conceptual grounding in the Web Service Modeling Ontology (WSMO) [31]. The Semantic Engine is supported by Web Services Execution Environment (WSMX) [32]. WSMX provides middleware functionality designed to take advantage of the semantic annotations of Web Services using the WSMO model. The implemented WSMX architecture provides an approach to the automated discovery, composition, mediation and invocation of Semantic Web Services. WSMX is intended as a middleware software layer at the endpoints of inter-service communications. This is an intent rather than a restriction. All information passed in and out of the WSMX boundary is represented in WSML. An adapter mechanism is provided to transform between non-WSML and WSML messages.

We propose a Learning Grid collaborative scenario with semantic capabilities (Figure 12) based on Learning Services of Sakai [33], WSML, WSMX, Globus Toolkit v4.07 and uPortal[34]. Sakai allows people to interact and work together as a distributed team. All the collaborative data in Sakai is maintained and can be

A Semantic Description Model

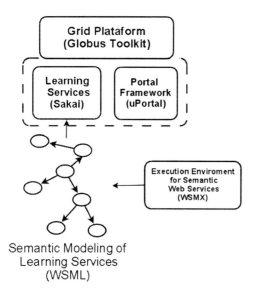

Fig. 12. Semantic Modeling of Learning Scenario

archived to associate the collaborative activity along with the results of any compute or experimental data associated with a particular research effort. uPortal is a portal framework composed of a set of Java classes and XML/XSL documents that allow to construct a front-end portal for an educational institutions.

4 Conclusions

In this work we have describe some approaches for constructing learning tools based on distributed resources that allow the consolidation of Learning Grid. Much of these efforts are focused on providing students, teachers and administrators with tools to enhance the interaction to each other and to facilitate the development of collaborative activities. In the same way an important issue in Learning Grid is the construction of scenarios depending on the user profile and preferences. A set of models to describe semantically Grid Collaborative Learning Scenarios were reviewed, which facilitates services automatic discovery and invocation without human interaction and enough information for human search. We comment some advantage of the evolution of a semantic representation based on OWL-S to one based on WSMO and to its posterior execution in an environment based on WSMX, which is centered on the representation of goals, pre-conditions and post-conditions of Grid Learning Services. The Learning Grid Portal model presented with semantic capabilities will allow the design of ubiquous and pervasive Collaborative Learning Scenarios, which will foment the interaction among the users and will allow developing collaborative and personalized learning activities.

References

1. Sherly, E., George, B., Deepa, L., Kumar, R.S.: A Grid Service Based Portal for Virtual Learning Campus (2004)
2. Chen, T.: Methodology for Supporting Novel Model of E-Learning Platform in Grid Architecture. In: Li, F., Zhao, J., Shih, T.K., Lau, R., Li, Q., McLeod, D. (eds.) ICWL 2008. LNCS, vol. 5145, pp. 314–321. Springer, Heidelberg (2008)
3. Luo, L., Fei, Y., Liang, J.: On Demand E-Learning with Service Grid Technologies (2004)
4. Ardaiz, O., Artigas, P., Díaz de Cerio, L., Freitag, F., Gallardo, A., Messeguer, R., Navarro, L., Royo, D., Sanjeevan, K.: ULabGrid, an Infrastructure to Develop Distant Laboratories for Undergrad Students over a Grid (2003)
5. Lischka, J., Karagiannis, D.: A Generic E-learning Engineering Framework Embracing the Semantic Web (2004)
6. Huang, C., Xu, F., Xu, X., Zheng, X.: Towards an Agent-Based Robust Collaborative Virtual Environment for E-Learning in the Service Grid (2006)
7. Tian, W., Mao, Y.: A Semantic Grid Application for E-Learning Data Sharing. In: Li, F., Zhao, J., Shih, T.K., Lau, R., Li, Q., McLeod, D. (eds.) ICWL 2008. LNCS, vol. 5145, pp. 457–467. Springer, Heidelberg (2008)
8. Zhuge, H., Li, Y., Bi, J., Cheung, T.-y.: KGCL: A Knowledge-Grid-Based Cooperative Learning Environment (2002)
9. Samaras, G., Karenos, K., Christodoulou, E.: A Grid Service Framework for Metadata Management in Self e-Learning Networks. The SeLeNe Consortium (2004)
10. Foster, I., Argonne, H.K., Fujitsu, A.S., Fujitsu: The Open Grid Services Architecture, V. 1.0, http://www.gridforum.org/documents/GWD-I-E/GFD-I.029v2.pdf
11. Ritrovato, P., Gaeta, M.: Towards a Semantic Grid for Human Learning, Centro di Ricerca in Matematica Pura ed Applicata (CRMPA), Dipartimento di Ingegneria dell'Informazione e Matematica Applicata (2005)
12. Gouardères, G., Conté, E., Mansour, S., Razmerita, L.: Ontology based user modeling for personalization of grid learning services (2005)
13. Li, M., Zhu, H., Zhu, Y.: PLANT: A Distributed Architecture for Personalized E-Learning. In: Leung, H., Li, F., Lau, R., Li, Q. (eds.) ICWL 2007. LNCS, vol. 4823, pp. 20–30. Springer, Heidelberg (2008)
14. Abbas, Z., Umer, M., Odeh, M., McClatchey, R., Ali, A., Ahmad, F.: A Semantic Grid-based E-Learning Framework (SELF), NUST Institute of Information Technology, CCCS Research Centre, University of the West of England
15. Guangzuo, C., Fei, C., Hu, C., Shufang, L.: OntoEdu: A Case Study of Ontology-based Educa-tion Grid System for E-Learning, Modern Education Technology Center at Peking Univ.
16. Liao, C.-J., Yang, F.-C.O.: A Workflow Framework for Pervasive Learning Objects Composition by Employing Grid Services Flow Language (2004)
17. Ivonne, R., Sara, P.: Extending and Augmenting Scientific Enquiry trough Pervasive Learning Environments. Children, Youth and Environments 14(2), 67–83 (retrieved), http://www.colorado.edu/journals/cye/
18. Vega-Gorgojo, G., Bote-Lorenzo, M.L., Gómez-Sánchez, E., Dimitriadis, Y.A., Asensio-Pérez, J.I.: Semantic Search of Learning Services in a Grid- Based Collaborative System. School of Telecommunications Engineering, Univ. of Valladolid

19. Goble, C., De Roure, D.: The Semantic Grid: Myth Busting and Bridge Building. In: Proceedings of the 16th European Conference on Artificial Intelligence (ECAI 2004), Valencia, Spain (2004)
20. Capuano, N., Gaeta, M., Ritrovato, P.: The Anatomy of the Learning Grid, The Learning Grid Handbook. IOS Press, Amsterdam (2008)
21. Di Martino, B.: An Ontology Matching Approach to Semantic Web Services Discovery. In: Min, G., Di Martino, B., Yang, L.T., Guo, M., Rünger, G. (eds.) ISPA Workshops 2006. LNCS, vol. 4331, pp. 550–558. Springer, Heidelberg (2006)
22. de Bruijn, J., Fensel, D., Kerrigan, M., Keller, U., Lausen, H., Scicluna, J.: Modeling Semantic Web Services - The Web Service Modeling Language. Springer, Heidelberg (2008)
23. IMS Global Learning Consortium, IMS Abstract Framework: White Paper (2003)
24. Gutiérrez-Carreón, G., Daradoumis, T., Jorba, J.: Semantic Description of Grid Based Learning Services. In: Min, G., Di Martino, B., Yang, L.T., Guo, M., Rünger, G. (eds.) ISPA Workshops 2006. LNCS, vol. 4331, pp. 509–518. Springer, Heidelberg (2006)
25. Dean, M., Schreiber, G., Bechhofer, S., van Harmelen, F., Hendler, J., Horrocks, I., McGuinness, D.L., Patel-Schneider, P.F., Stein, L.A.: OWL Web Ontology Language Reference (2004).
26. Srinivasan, N., Paolucci, M.: Katia Sycara, Semantic Web Service Discovery in the OWL-S IDE (2006)
27. Kashyap, V., Bussler, C., Moran, M.: The Semantic Web - Semantics for Data and Services on the Web. Springer, Heidelberg (2008)
28. de Bruijn, J., Fensel, D., Kerrigan, M., Keller, U., Lausen, H., Scicluna, J.: Modeling Semantic Web Services -The Web Service Modeling Language, 4th edn. Springer, Heidelberg (2008)
29. Lytras, M.D., Naeve, A.: Ubiquitous and Pervasive Knowledge and Learning Management: Semantics. In: Social Networking and New Media to Their Full Potential. Idea Group Publishing (2007)
30. The Web Service Modeling Language (WSML), http://www.wsmo.org/wsml/wsml-syntax
31. Web Service Modeling Ontology (WSMO), http://www.wsmo.org/TR/d2/v1.2/
32. Maximilian Herold, WSMX Documentation, Digital Enterprise Research Institute Galway (2008)
33. Sakai Project, http://sakaiproject.org
34. Chamberlain, L., Juahkah, M.N., Sherratt, R.: Beginners Guide to uPortal (2003)

Author Index

Anguita-Martínez, Rocío 129

Caballe, Santi 65, 113
Capuano, Nicola 53
Casillas, Luis 99
Charalambos, Vrasidas 19
Constantinou, Charalambos 19

Daradoumis, Thanasis 83, 99
Demetriadis, Stavros 1
Dimitriadis, Yannis 129

Feldman, Jerome 65, 113

Gutiérrez-Carreón, Gustavo 171

Jorba, Josep 171
Juan, Angel A. 147

Karakostas, Anastasios 1
Kordaki, Maria 37, 83

Lázaro, Daniel 147
Li, Joseph 65

Marcos-García, José Antonio 129
Marquès, Joan Manuel 147
Martínez-Monés, Alejandra 129
Miranda, Sergio 53

Orciuoli, Francesco 53

Papadopoulos, George 19

Retalis, Symeon 19
Rubia-Avi, Bartolomé 129
Ruiz-Requies, Inés 129

Thaw, David 65, 113

Vilajosana, Xavier 147

LaVergne, TN USA
25 September 2009
159086LV00001B/109/P

9 783642 040009